Integrated Water Resource Management in Brazil

Integrated Water Resource Management in Brazil

Editors:

Carsten Lorz, Franz Makeschin and Holger Weiss

Published by   **IWA Publishing**
Republic - Export Building
1 Clove Crescent
London E14 2BA, UK
Telephone: +44 (0)20 7654 5500
Email: publications@iwap.co.uk
Web: www.iwaponline.com

First published 2014
© 2014 IWA Publishing

Disclaimer
The information provided and the opinions given in this publication are not necessarily those of IWA and should not be acted upon without independent consideration and professional advice. IWA and the Author will not accept responsibility for any loss or damage suffered by any person acting or refraining from acting upon any material contained in this publication.

British Library Cataloguing in Publication Data
A CIP catalogue record for this book is available from the British Library

ISBN 9781789065350 (Paperback)
ISBN 9781780404905 (eBook)

Contents

Chapter 3

Chapter 4

Chapter 5

Chapter 6
Water quality of tropical reservoirs in a changing world – the case of Lake Paranoá, Brasília, Brazil
G. Abbt-Braun, H. Börnick, C. C. S. Brandão, C. B. G. Cavalcanti,
C. P. Cavalcanti, F. H. Frimmel, M. Majewsky, B. Steiniger, M. Tröster and E. Worch

Chapter 7
Bridging the gap: Current and future drinking water treatment for a fast-growing megacity – Brasília, Distrito Federal
E. Vasyukova, K. R. F. O. Dassan, F. Braga, C. Simões, T. Baylão, K. Neder and W. Uhl

Chapter 8
Developing the urban water system towards using the Paranoá Lake in Brasília as receptor and water resource 109

F. W. Günthert, V. Freitas, K. Neder, A. Obermayer, S. Faltermaier
and C. Tocha

Chapter 9
Capacity development as base element of Integrated Water Resource Management in Central Brazil 125

B. Kuse, J. Wummel and K. D. Neder

Chapter 1

Introduction

C. Lorz, F. Frimmel, W. Günthert, S. Koide, F. Makeschin, K. Neder,
W. Uhl, D. Walde, H. Weiss and E. Worch

1.1 INTEGRATED WATER RESOURCE MANAGEMENT IN CENTRAL BRAZIL

The demand for concepts of integrated water resource management (IWRM) is increasing worldwide due to rising awareness for sustainable use of water resources and due to pressure from society for reliable and healthy water supply. For many regions in South America, where water scarcity and/or water quality are issues, this applies in particular. In addition to the specific natural conditions, that is, strong seasonal contrasts and strong climatic variability, the rapid changes of land use/cover – mostly caused by the drastic expansion of agricultural land and urbanization processes – have severe effects on water resources. Effects of land use change on water resources seem to exceed the effects of climate change, but global climate models predict substantial changes in future climate and in consequence, severe effects on water resources are to be expected. As a result of the rapid changes of land use, rising demand for water supply and rising production of waste water can be expected – both in terms of amount and spatial expansion – due to higher population densities caused by natural population growth and migration as well as higher per capita consumption. The understanding of the complex interactions between water, climate, land use, society and water technologies is a crucial step to achieve sustainable water supply with high standards in regard of quality and reliability.

Central Brazil belongs to the type of region described above. As climate change and very dynamic processes of urbanization and expansion of agriculture are happening in this area, substantial impacts on water resources have been observed. Urbanization and high shares of urban population as well as substantial expansion of croplands are seen as major causes for ecological problems in Brazil, for example, not sustainable use and pollution of water resources (Braga *et al.* 2008; Hespanhol, 2008; Tucci, 2001, 2008). The adaptation to changing frame conditions as well as the prevention and mitigation of negative impacts on water resources are the main challenge for the sustainable water supply of the region. Therefore, a consortium of Brazilian and German partners decided to start the IWRM project IWAS-ÁGUA DF[1] focussing on the Distrito Federal including the national capital Brasília.

1.2 THE PROJECT IWAS-ÁGUA DF

During the period 2008–2013 the project IWAS-ÁGUA DF was carried out as joint activity by Brazilian and German partners, aiming at providing the scientific base for the sustainable use of water resources within the Distrito Federal (DF). The study area was chosen because of its model character due to outstanding data base on water relevant information and the existing high standards in terms of technology and capacity development in the water sector.

The general objective of the project was to contribute to the development of an IWRM approach for the Distrito Federal, identifying causes of problems and proposing solutions to maintain water supply for the region also with a focus on

[1]IWAS-ÁQUA DF was funded through the program 'International Water Research Alliance Saxony (IWAS) – Management of Water Resources in Hydrological Sensitive World Regions' by the BMBF (FKZ 02WM1165/66 and 02WM1070) and by Brazilian Partners. For more information go to http://www.ufz.de/iwas-sachsen/index.php?en=18049

changing frame conditions. The project follows the DPSIR approach (EEA, 1999 [Figure 1.1]), which identifies the driving forces and pressures on the compartments river basins, drinking water and waste water systems and analyses the state, impacts and response of these compartments. In effect, this results in a subdivision into three major complexes – which represents at the same time the structure of the book – (a) river basins and water bodies (Chapter 2–5), (b) surface/drinking/waste water (Chapter 6–8) and (c) capacity development (Chapter 9).

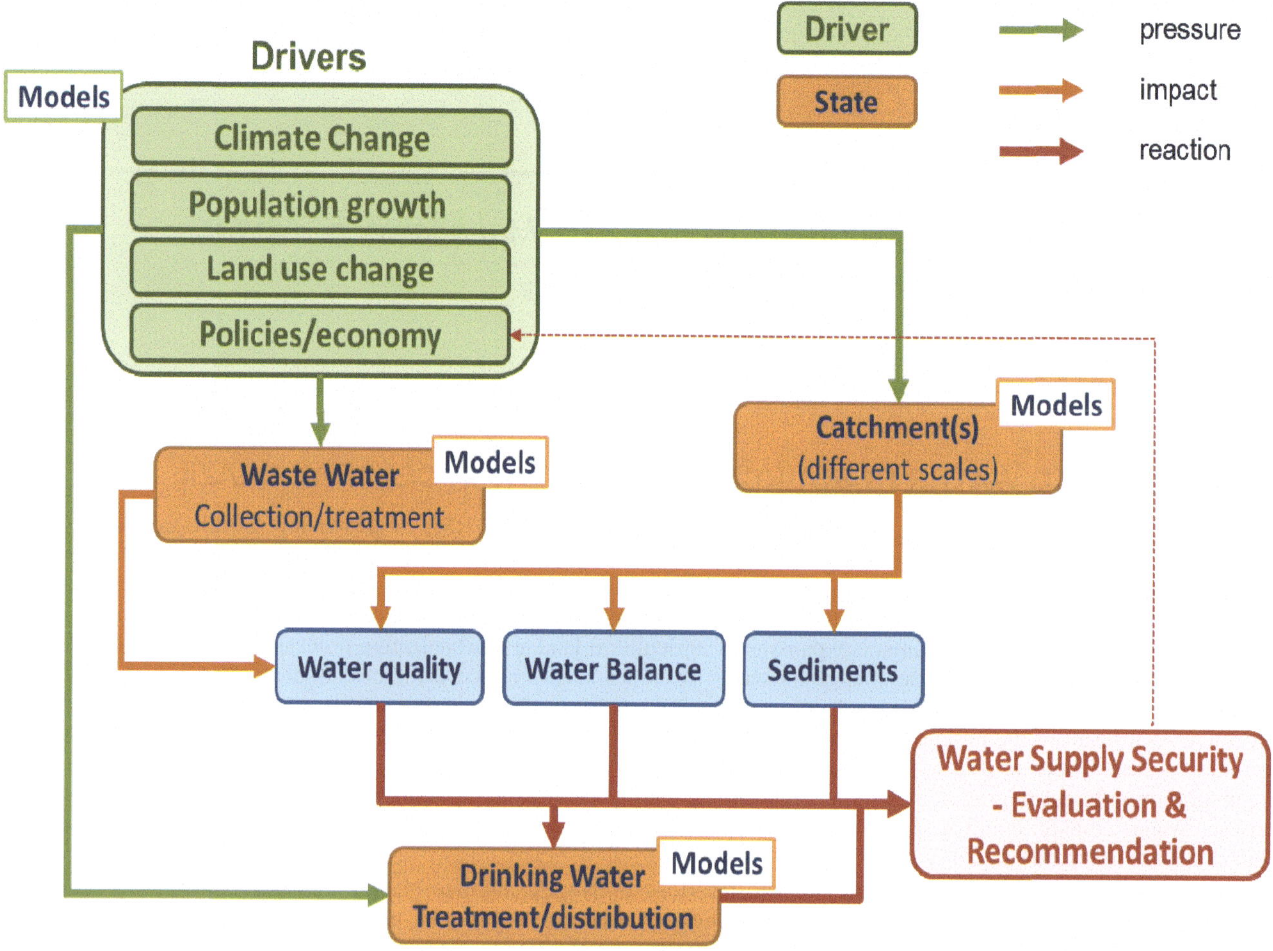

Figure 1.1 Driving forces-Pressures-State-Impacts-Responses (DPSIR) scheme of the project (modified after Lorz *et al.* 2012).

1.3 THE STUDY REGION

1.3.1 Climate

The Distrito Federal covers an area of 5790 km^2 and is located in the high plains (*planalto*) of Western Central Brazil with altitudes between 1000 and 1450 m a.s.l. The topography might be characterized as rolling landscape where steep slopes are rather an exception. The geological underground consists of series of metasediments, that is, argillic to sandy rocks of the Palaeozoic period. Soils are mostly Oxisols, Cambisols and Gleisols developed in the regolith mantles of the rocks mentioned above.

The study region belongs to the outer Tropics (Aw climate after Köppen) having mean annual precipitation of 1300–1700 mm and mean annual temperatures of 20–21°C (WMO, 2010). The climatic conditions are characterized by strong seasonality. The dry season comprises 5–6 months, during south-winter, from late March to late September. The average amount of precipitation during rainy season is four times higher than during dry season. However, the temporal variability of mean annual temperature and precipitation as well as in terms of seasonality is high.

Effects of climate change are already observable for the last three decades (Borges *et al.* 2013). Since the end of the 1970s a trend to longer dry periods with less rain days at the begin and the end of the dry seasons as well as a general trend for lower mean annual precipitation have been observed for five stations in the Distrito Federal (Table 1.1). However, this trend is not significant for all stations and will not explain fully the dramatic decrease of base flow discharge during the dry period (Lorz *et al.* 2012). For the future, a further increase of seasonality, that is, loner dry seasons, is predicted (Borges *et al.* 2013)

Table 1.1 Mean annual precipitation for five stations (see Figure1.2 for location) in the Distrito Federal and trends (Kendall´s tau) for rain days and annual precipitation (Mann-Kendall test) (Lorz *et al.* 2012).

Monitoring station	Period	MAP (mm)	Rain days April/May	Rain days October	Annual precipitation
A 1547013	1978–2009	1332	-0.25^{+} *(May)*	**−0.26***	-0.24^{+}
B 1547014	1979–2006	1480	**−0.36** (April)**	-0.18^{+}	0.08^{+}
C 1547015	1978–2004	1419	-0.17^{+} *(May)*	**−0.26***	-0.04^{+}
D 1548007	1978–2008	1560	-0.25^{+} *(May)*	**−0.26***	-0.12^{+}
E 1548008	1979–2006	1444	**−0.31* (May)**	**−0.35****	**−0.33***

[+] = two-sided p > 0.05 (not significant), * = two-sided p < 0.05 (significant), ** = two-sided p< 0.01 (very significant), rain days are days with any recorded rainfall.

1.3.2 Land use

The region is part of the Cerrado biome, which covers nearly a quarter of the total surface area of Brazil, that is, 2.5 million km^2. The dominating natural vegetation form is savannah (Figure 1.2). Grass dominated savannah (*Campo*) is interchanging with typical tree savannah (*Cerrado*, Figure 1.4a) depending on topography and water availability (Oliveira & Marquis, 2002; Silva *et al.* 2006).

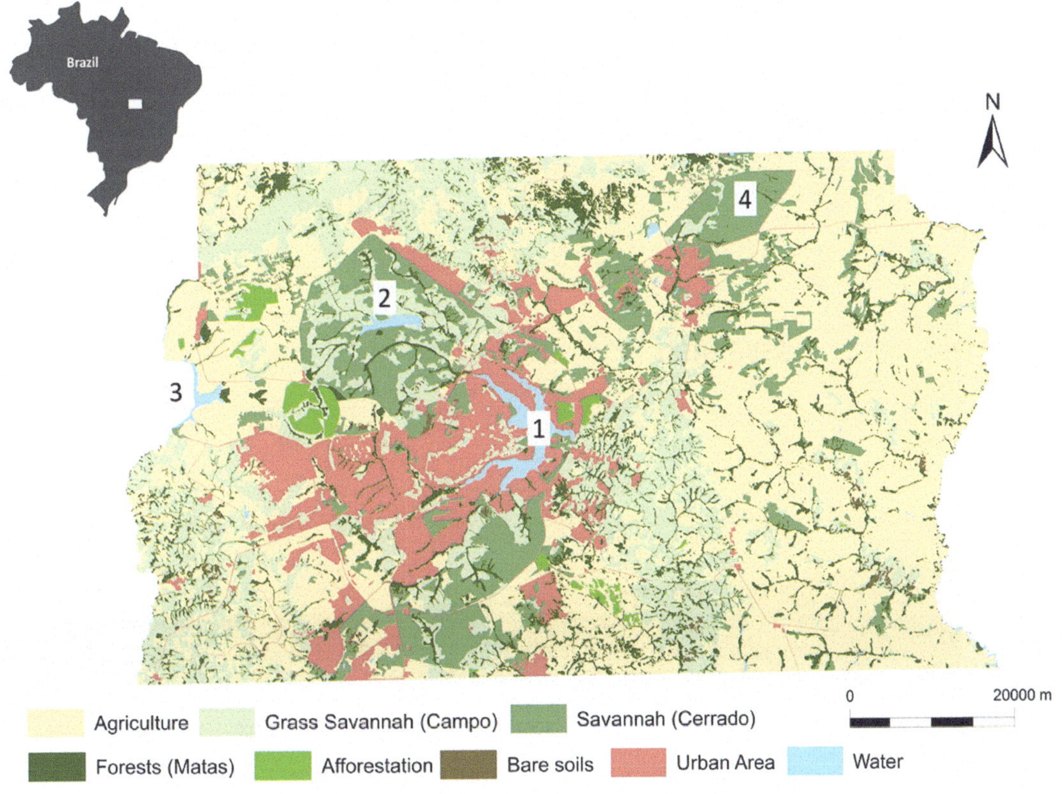

Figure 1.2 Location and land cover of the Distrito Federal in 2006, 1 = Lago Paranoá, 2 = Lago Sta. Maria, 3 = Lago Descoberto, 4 = Pipiripau (after Fortes *et al.* 2007).

Forested areas are divided in dry forests (*Cerradão*), gallery forests (*Mata de Galeria*) and pine or eucalyptus plantations. Protected areas include the National Park of Brasília – around the reservoir Santa Maria (Fig. 1.2) – and several smaller state reserves. Most of the region is covered by pastures and arable fields. Large mechanized agriculture cultivating soy bean (Figure 1.4b), corn and beans account for more than 90% of the total arable land (IGBE, 2010). Large scale crop cultivation with no-tillage practice is prevailing in the eastern part of Distrito Federal, whereas in the western part small scale farming and horticulture are dominant (Figure 1.4c). Large areas are urbanized with all development levels, that is, irregular settlements with less developed infrastructure, highly developed urban areas and residential areas with high development standards. Industrial areas do not exist in the region.

The region has experienced a substantial change in land cover and land use since the foundation of the national Capital Brasília in 1960, as it has been also observed for other regions in Brazil (Simon *et al.* 2010). Land use/cover has changed mostly due to a substantial increase of agricultural land and urban sprawl occupying mostly natural areas, that is, savannah or forest (Figure 1.3). The loss of areas with natural vegetation has been quantified with 58% for the period 1954–1998 in DF (UNESCO, 2002) which is in the same order as reported for the Cerrado biome (Klink & Machedo, 2005). The share of (semi)natural vegetation was around 40% in 2006 for the DF. From 2002 to 2007 the area of arable land (without orchards) increased by 47%, from 84,240 ha in 2002 to 123,692 ha in 2007.

Urbanization is the second major process of land use change for the Distrito Federal and is associated with urban sprawl and the spatial expansion of sealed areas. The share of urban areas increased for the period 1954 to 2001 from 0.02% to 10.62% (CODEPLAN, 2007; Fortes, 2007). The nucleus of Brasilia – finished in 1960 (Figure 1.4d, e) – is the so called *Plano Piloto* where the outlines have the shape of an airplane (in Figure 1.4d the central axis represents the fuselage). Since then, the settlement structure developed polycentric with fast growing suburbs, for example, Itapoã (Figure 1.4f). Recently, new upper standard residential areas have been developed, for example, the quarter Noroeste (Figure 1.4g).

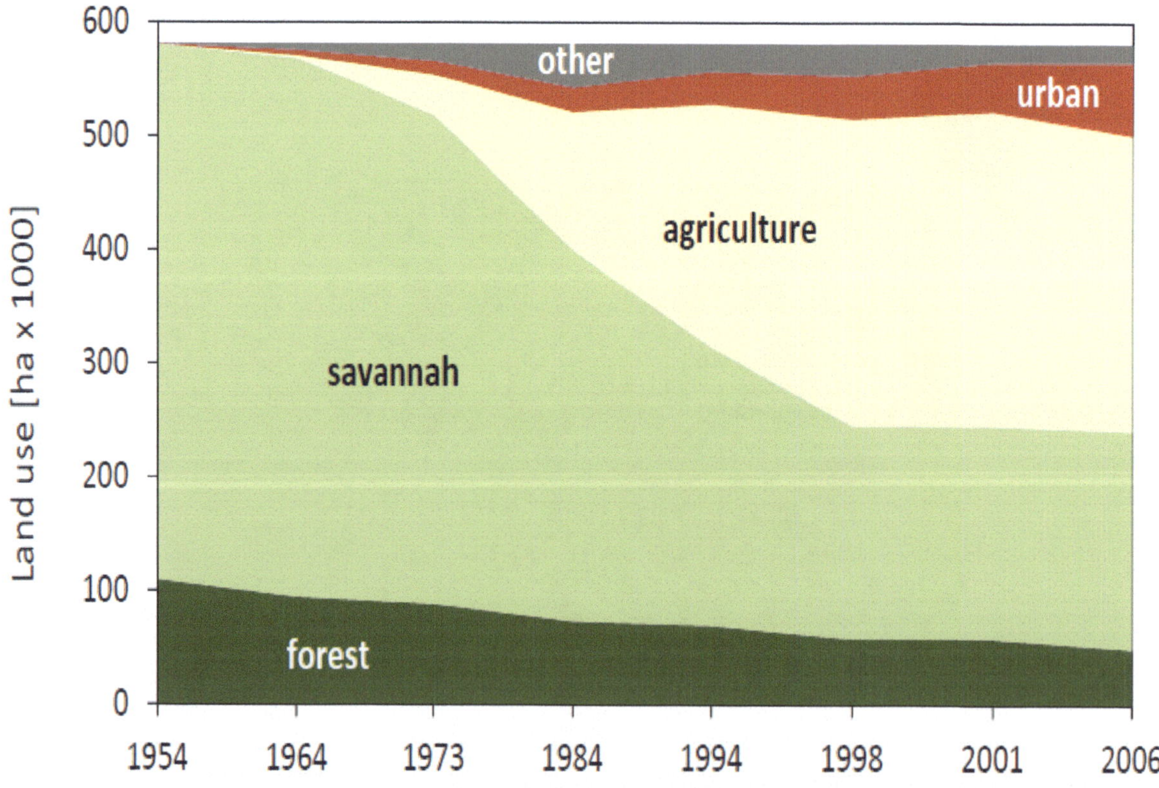

Figure 1.3 Land use change from 1954 to 2006 in the Distrito Federal (Fortes *et al.* 2007 and UNESCO, 2002).

Figure 1.4 (a) tree savannah (*Cerrado*) in the national park of Brasília, (b) large scale agriculture (soy bean) in the northeastern DF, (c) irrigated strawberry fields in the western DF, (d) the central east-west axis (*Eixo Monumental*) of Brasília with buildings of the federal government and the Lago Paranoá in the left background, (e) view from southwest on Brasília, (f) commercial area in Itapoã, (g) new built residential areas in NW-Brasília (*Noroeste*), (h) water extraction site in Santa Maria reservoir.

1.3.3 Demographic development and water supply

The Brasília region is one of the large urban aggregations of Brazil where the capacities of the existing systems for water supply are already near their limits and demand will increase dramatically in the near future (ANA, 2009; CAESB, 2010). For the Distrito Federal the situation is even more pressing, since 94% of the population live in urban areas. The predicted growth of population for the region from currently 2.5 million to more than 3.2 million in 2025 will mostly take place in urban areas (Figure 1.5). Estimations of future water demand predict an exceeding of available water system capacities for the near future. During dry season the capacities for some drinking water treatment plants are already close to maximum. The average consumption per capita was in 2003 approximately 195 l d^{-1}, but minimum (110 l d^{-1}) and maximum average (597 l d^{-1}) per municipality are far apart (PGIRH, 2006).

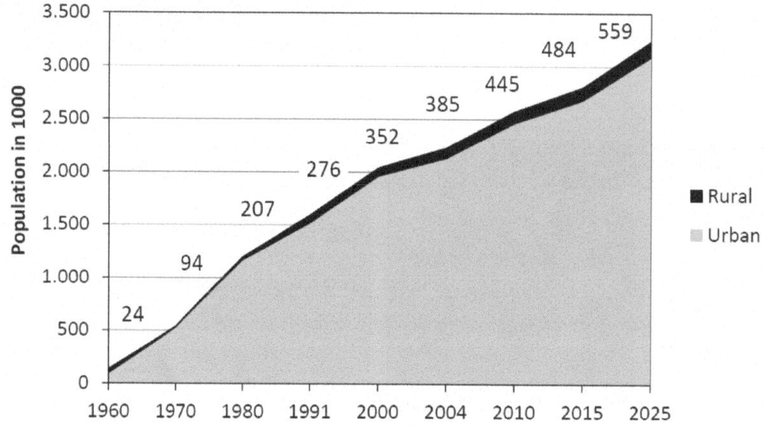

Figure 1.5 Historic and predicted population of the DF from 1960 to 2025; numbers above columns are population density (inhabitants per km^2), values for 2015 and 2025 are estimated (PGIRH, 2006).

Currently, water supply for the region is provided by two major reservoirs (Figure 1.2), Lago Santa Maria and Lago Descoberto (Table 1.2). These reservoirs cover about 78% of the total water supply of the Distrito Federal. The remaining water comes from the extraction of stream water and of groundwater. For the future, the regional water supplier CAESB is planning to extract around 2.8 m^3 s^{-1} from the Lago Paranoá, an urban lake in the centre of Brasília. An additional plan in realization is to use the reservoir Corumba IV, a reservoir for hydropower generation around 80 km to the south, for the water supply of DF. However, both lakes might have water quality problems because the respective watersheds have less adequate land use and receive waste water from settlements.

Table 1.2 Contribution to the water supply of different types of water resources (CAESB, 2003).

Type of water resource	Discharge (m^3 s^{-1})	Percentage (%)
Lago Santa Maria (area: 6 km^2, volume: 58 * 10^6 m^3)	1.9	21
Lago Descoberto (area: 14.8 km^2, volume: 102 * 10^6 m^3)	5.1	57
Stream water	1.6	18
Groundwater	0.3	4
Total water supply	8.9	100
Lago Paranoá (area: 38 km^2, volume: 498 * 10^6 m^3)	2.8	(29)

1.4 REFERENCES

ANA (Agência Nacional de Águas) (2009). Atlas Regiões Metropolitanas: Abastecimento Urbano de Água: Resumo Executivo. Agência Nacional de Águas, Brasília.

Borges P. A., Franke J., Silva F. D. S., Weiss H. and Bernhofer C. (2013). Differences between two climatological periods (2001–2010 vs. 1971–2000) and trend analysis of temperature and precipitation in Central Brazil. *Theoretical and Applied Climatology*, doi: 10.1007/s00704-013-0947-4.

Braga B. P. F., Flecha R., Pena D. S. and Kelman J. (2008). Federal pact and water management. *Estudos Avançados*, **22**(63), 17–41.

CAESB (Companhia de SaneamentoAmbiental do Distrito Federal) (2010). Internal Report (unpublished).

CODEPLAN (Companhia de Planejamento do Distrito Federal) (2007). Plano Director de Ordenamento Territorial do Distrito Federal. Governo do Distrito Federal, Brasilia.

EEA (European Environment Agency) (1999). Environmental indicators: typology and overview. *Technical Report*, **25**, 1–19.

Fortes P. T., Oliveira G. I., Crepani E. and Medeiros J. S. (2007). Geoprocessamento aplicado ao planejamento e gestãoambiental na Área de Proteção Ambiental de Cafuringa, Distrito Federal Parte 1: processamento digital de imagens. In *Anais XIII Simpósio Brasileiro de Sensoriamento Remoto*, Florianópolis, Brasil: 2605–2612.

Hespanhol I. (2008). A new paradigm for water resource management. *Estudos Avançados*, **22**(63), 131–157.

IBGE (InstitutoBrasileiro de Geografia e Estatística) (2010). Pesquisa Nacional por Amostra de Domicílios. http://www.ibge.gov.br (accessed 23 June 2010).

Klink C. A. and Machedo R. B. (2005). A conservação do Cerrado brasileiro. *Megadiversidade*, **1**(1), 147–155.

Lorz C., Neumann C., Bakker F., Pietzsch K., Weiss H. and Makeschin F. (2013). A web-based planning support tool for sediment management in a meso-scale river basin in Western Central Brazil. *Journal Environmental Management*, DOI 10.1016/j.jenvman.2012.11.005.

Lorz C., Abbt-Braun G., Bakker F., Borges P., Börnicke H., Frimmel F., Gaffron A., Höfer R., Makeschin F., Neder K., Roig L. H., Steiniger B., Strauch M., Worch E. and Weiss H. (2012). Challenges of an Integrated Water Resource Management for the DF of Brasília, Western Central Brazil – climate, land use and water resources. *Journal of Environmental Earth Sciences*, **65**(5), 1601–1611.

Oliveira P. S. and Marquis R. J. (2002). The Cerrados of Brazil. Columbia University Press, New York.

PGIRH (2006). Plano de gerenciamentointegrado de recursoshídricos do Distrito Federal PGIRH/DF 1-6., GDF Brasília, Secretaria de Infra-Estrutura e Obras.

Silva J. F., Farinas M. R., Felfili J. M. and Klink C. A. (2006). Spatial heterogeneity, land use and conservation in the Cerrado region of Brazil. *Journal of Biogeography*, **33**, 536–548.

Tucci C. E. M. (2001). Some scientific challenges in the development of South America's water resources. *Hydrological Sciences*, **46**(6), 937–946.

Tucci C. E. M. (20sssssssssssp08). Urban waters. *Estudos Avançados*, **22**(63), 97–111.

UNECSO (United Nations Educational, Scientific and Cultural Organization) (2002). Vegetação o Distrito Federal, tempo e espaço, uma avaliação multitemporal da perda de cobertura vegetal no DF e da diversidadeflorística da reserva da biosfera do cerrado – Fase I. UNESCO, Brasília .

WMO (World Meteorological Organization) (2010). World Weather Information Service: Brasília. http://www.worldweather.org/136/c00290.htm (accessed 15 January 2010).

Chapter 2

Climate change in Central Brazil[1]

P. Borges, F. D. dos Santos Silva, H. Weiss and C. Bernhofer

2.1 INTRODUCTION

The climate of the DF shows the characteristics of Central West climate, with hot and rainy summers, and dry and mild winter. Considering Köppen's classification, the region is classified as Cwa and Cwb climate. Some authors (Alves, 2009; Maia & Baptista, 2010) classify the climate as tropical wet and dry climate or Savanna climate (Aw after Köppen's classification). This climate is characterized by particular temporal distribution of temperatures and precipitation with two very well defined seasons with dry and cold winters and warm and humid summers (Alves, 2009). In average, the mean surface air temperature is 20.5°C, while rainfall amounts to 1500 mm per year (Ramos *et al.* 2009). However, more than 90% of the precipitation occurs from October to April and potential evaporation has its annual maximum in the driest months of the year, that is, June, July and August which are the months with lowest precipitation in Brasília. The high amount of annual rainfall, the heterogeneous distribution over the year and the high evaporation rates can strongly affect the water availability, especially in a water supply system which depends on surface water from large reservoirs. Nevertheless, changes in the regional climate are expected for the 21th century. The Brazilian National Water Agency (ANA, 2005) projects that the semi-arid and savanna regions in Brazil, that is, Cerrado, may face water supply crises in the coming decades. Studies on future climate scenarios of the Brazilian National Institute for Space Research (INPE; Ambrizzi *et al.* 2007) demonstrates that the Central West region may face an increase in temperature of 2 to 6°C until the end of this century. Moreover, the same study predicts a substantial increase in extreme rainfall events and droughts. Marengo (2007) describes possible impacts of climate change, for instance increase in evaporation rate and heat waves which may likely affect public health, agriculture production and hydropower generation.

Facing the urgency to take action that will guarantee the water supply of Brasília, the project IWAS-ÁGUA DF aims to contribute to the development of an Integrated Water Resources Management (IWRM) system in the region. In order to achieve this goal, the project is organized in a toolbox concept wherein climate is assumed an essential input for impact modeling (Lorz *et al.* 2012).

A systematic analysis of observed climatic variability contributes to a better understanding of the components of the regional climate in a region. However, available historical and current climate studies in Central Brazil did not provide the necessary spatial and temporal resolution for detailed climate diagnosis, as they have not satisfied the requirements for developing regional climate scenarios (Borges *et al.* 2014a). Figure 2.1 shows the general steps within the project IWAS-ÁGUA DF adopted here for developing climate scenarios. In a first step, key features of the present climate regime (i.e., Baseline) such as seasonality, spatial distribution and trends, as well as data quality, were investigated. Furthermore, GCMs can provide valuable information about climate change on a regional scale (Christensen *et al.* 2007; van der Linden & Mitchell, 2009), and therefore the output of all Global Circulation Models (GCMs) used in the Fourth Assessment Report

---

[1]Thanks to the international modeling groups for providing their data, the Program for Climate Model Diagnosis and Intercomparison (PCMDI) for collecting and archiving the data, the Coupled Model Intercomparison Project (CMIP) and for organizing the model data analysis, and technical support IPCC WG1 TSU. Access to data and technical assistance is provided by the Model and Data Group (M & D) of the Max-Planck-Institute for Meteorology. Special gratitude to Dr. Lauro Tadeu Guimarães Fortes and collaborators of the Coordenação-Geral de Desenvolvimento e Pesquisa – CDP of INMET for supporting this work and providing all assistance needed.

(AR4) of the Intergovernmental Panel on Climate Change (IPCC) from the Coupled Model Intercomparison Project (CMIP3) project were summarized using the multi-model ensemble approach. Finally, as users may require fine-scale or local data about climate variability (Mearns *et al.* 1999; Lu, 2006), the capacity of a Statistical Downscaling Method (SDSM after Wilby *et al.* 2002) in providing regional climate scenarios for impact assessment is demonstrated.

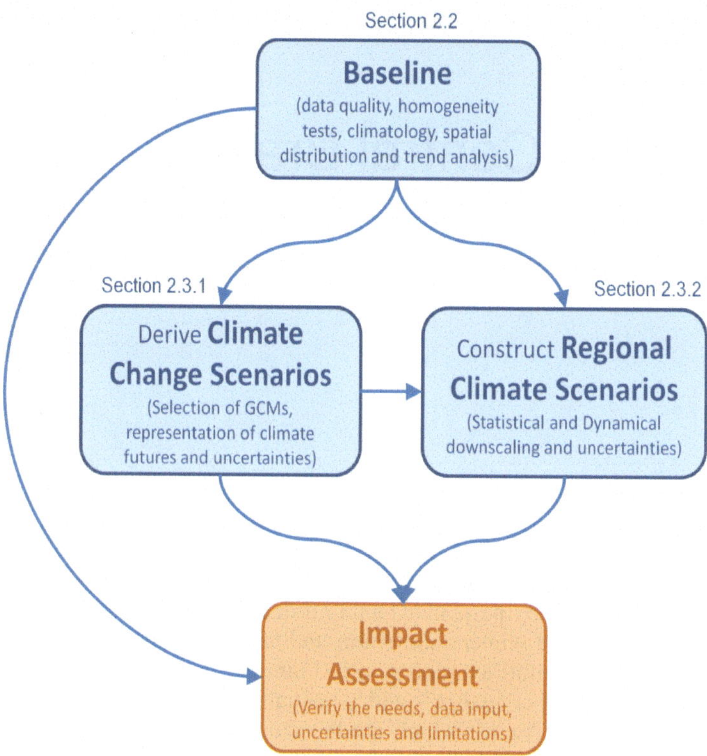

Figure 2.1 Workflow for the development of climate scenarios for impact assessment, with references to the sub-sections where the steps are demonstrated.

2.2 BASELINE CLIMATE

The recent climate condition is essential information to assess implications of future changes in the water resources. According to Lu (2006), '*baseline climate is the climatic conditions that are representative of present day or recent prevailing climatic trends for a given period of time in a specific geographic area*'. The baseline describes average conditions, as well as the spatial and temporal variability of climate variables of interest. The baseline has four objectives, (1) give information for a better understanding of features and patterns of present climate; (2) identify likely ongoing trends; (3) provide a benchmark for comparisons against future changes; and (4) calibrate and validate the performance of climate models. Nevertheless, observational data of high good quality is crucial for defining a reliable baseline climate (IPCC-TGCIA 1999).

2.2.1 Regional climate

Based on climatological normals of 1961–1990 compiled by the Brazilian National Institute of Meteorology – INMET (Ramos *et al.* 2009), the rainy season runs from October until April. During this period, the strong warming of the atmosphere and the release of latent heat generate a system at high atmospheric levels, known as the Bolivian High. Conversely, in low levels, a large area of low atmospheric pressures favors the occurrence of rainfall in the Central-West and consequently, in the DF. Monthly average can reach around 250 mm in December and January, and annual rainfall is around 1500 mm. The average temperature is 22°C in the period from October to April. During dry season, between May and September, mean temperatures observed are around 18-20°C. Although the maximum temperatures remain between 24 and 26°C, the low relative humidity cause strong sensation of heat in the afternoons, and cold during the nights and early mornings. This period is influenced by a strong dry continental air mass, which leaves the sky with few clouds throughout the period, increasing visibility, but favouring the occurrence of strong temperature inversions near the surface. The combination of dryness and low dissipation of particulates in the air can considerably aggravate the air quality in Brasília. There are approximately 2400 hours of sunshine per year, with July showing an average maximum of 266 hours of sunshine, and December a minimum average of 138 hours of sunshine. The relative humidity shows an annual average of 67%. December is the most humid month, with an average of 79%, while August the driest month with an average of 49%. However, values around 10% are common and may occur at certain times in the driest months of the year, that is, between June and September.

The average wind speed is 2.6 m/s, the values may slightly vary from month to month. The predominant wind direction is easterly from April to September. In the rainy season, northwest winds are more common. November and December are the months with highest cloudiness, 0.8 tenths, while June to August, is the lowest rates of cloudiness, 0.3 tenths. The mean annual average is 0.6 tenths. The mean annual air pressure is about 886 hPa, with the driest months having the highest values, around 889 hPa, and the rainiest months the lowest values, around 884 hPa.

2.2.2 Observations

2.2.2.1 Database (CLIMA-DF)

In the framework of the IWAS approach (Lorz *et al.* 2012), the working group Climate developed a databank called CLIMA-DF (Borges *et al.* 2014a). It comprises daily values of five climate elements: surface air temperature [°C] (i.e., mean and maximum), accumulated precipitation [mm], relative humidity [%], wind velocity [m/s] and global radiation [j/cm^2] (derived from insolation observations). Except of precipitation, most of the data were provided by the Brazilian National Institute of Meteorology (INMET). Precipitation datasets were mostly obtained from the Hydrological Information System HIDROWEB run by Brazilian National Water Agency (ANA). Additional datasets were provided by the Brazilian Enterprise for Agricultural Research (EMBRAPA) and Environmental Sanitation Company of the DF (CAESB). The databank includes time series of 37 meteorological stations and 120 rain gauges from Central Brazil (14 to 18° south; 44 to 51° west, see Figure 2.2). Precipitation datasets have recordings starting from 1941. However, most of the registered observations started only in the 1970s. For all other climate elements datasets start not earlier than 1961. Concerning the distribution of the observation network over space, most of the stations are concentrated in the DF area. In contrast, areas northern than latitude -15° (Figure 2.2) shows a very low observation density. Datasets were selected on the basis of length of record, data completeness, and spatial distribution across the region. Borges *et al.* (2014a) used 5 meteorological stations and 55 rain gauges for the visualization of spatial distribution climatological periods (1971–2000 and 2001–2010) and analysis of trends.

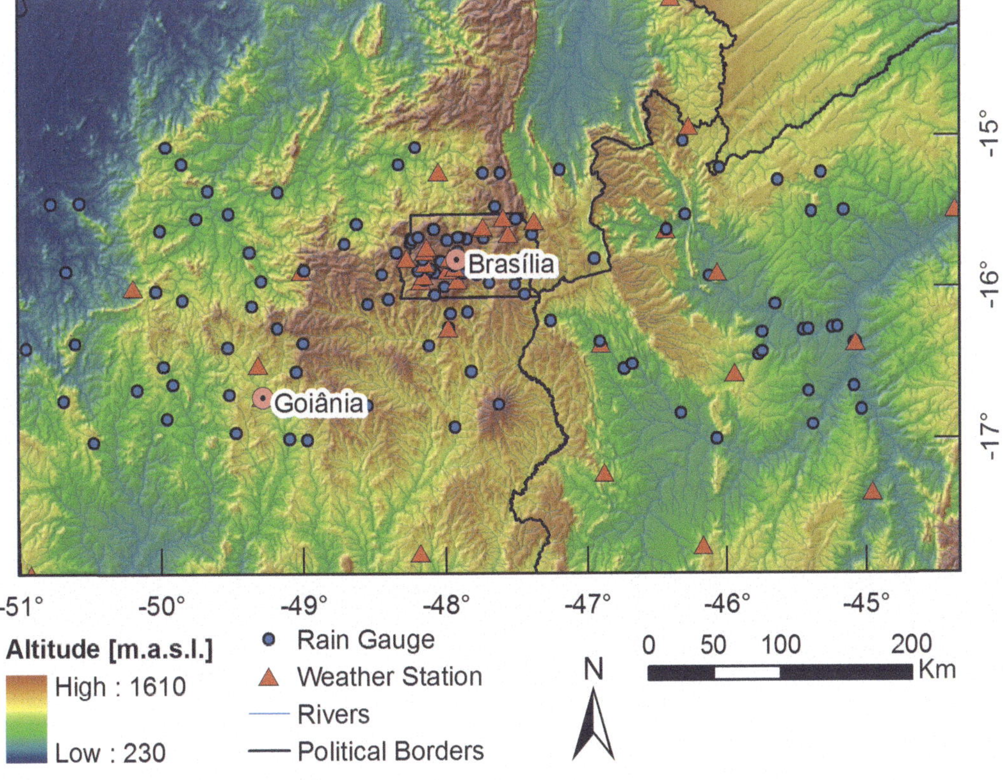

Figure 2.2 Spatial distribution of the observation network in Central Brazil.

Long-term climate time series will be affected by several non-climatic factors which might lead to misinterpretations due to changes in the location of an observation site, changes in the instrumentation, changes in procedures in observations or process data and changes in the local environment around an observation site (Trewin, 2010). Some changes cause distinct discontinuities while other changes, particularly change in the environment around the station such as urbanization, can cause gradual biases in the data. The CLIMA-DF databank comprises data from different source institutions and all of them apply their own quality control system. Observation methods and data quality control implemented by INMET follow the international standards suggested by the World Meteorological Organization (WMO, INMET 1999). HIDROWEB also controls the input of precipitation datasets in its databank using a consistency test tool called 'Hidro-PLU'. If any suspicious value is found, documentations are checked and station operators are consulted. However, a reliable Climate Atlas and trustworthy statistical analysis for the Central Brazil is desired and therefore a customized approach, based on the *Statistische Untersuchungen regionaler Klimatrends in Sachsen* – CLISAX (Bernhofer *et al.* 2008), was designed for the CLIMA-DF database.

The tool is able to check for suspicious values, errors and outliers by a 1- or 2-tailed test after Dixon (1950) and the homogeneity of the data is represented by graphical, that is, Craddock test, double sum analysis, quotient criteria and difference in limits, and numerical tests, that is, Abbe, Buishand and Standard Normal Homogeneity Test – SNHT (Craddock, 1956; Buishand, 1982; Alexandersson, 1986; Dahmen & Hall 1990; Herzog & Müller-Westermeier, 1998). Climatological periods, that is, 1971–2000 and 2001–2010, were calculated according to the suggestions of WMO (1989). Borges *et al.* (2014b) analyzed the performance of several spatial interpolation methods for seasonal and annual long-term mean precipitation and mean temperature in Central Brazil. The multivariate regression model using altitude, latitude and longitude as explanatory variables, using interpolation of residuals by inverse distance weighting (IDW), performed the most reliable and detailed predictions according to visual analysis and statistical criteria. Since trends are of high interest for climate impact studies, Borges *et al.* 2014a investigate the variability of mean surface air temperature, as well as precipitation, at seasonal and annual time scale in Central Brazil. Trends were analyzed by applying Rapp (2000), for calculation of the slope, and Mann-Kendall (Mann, 1945; Kendall, 1970) method to assess the significance of the trends.

2.2.2.2 Climatology 1971–2000 and spatial distribution

The spatial and temporal distribution of climate variables can be explained by the dependence of Central Brazil's climate on synoptic and mesoscale climate drivers such as tropical, extra tropical atmospheric circulations, and topography (Alves, 2009). Borges et al. (2014a) demonstrated that the spatial distribution of mean surface air temperature depends mainly on geographic position and topography. A topographic gradient in annual mean temperature is about -0.7°C/100 meters. The annual mean temperature for the normal climatological period 1971–2000 is 22.2°C and varies between 19.4°C and 24.5°C. The warmest regions are the lowlands in the north-west, the north-east and the south-east. While the northern highlands in DF and Goiás plateaus have milder temperatures. The warmest season registered was spring (SON), with average temperature of 23.2°C, followed by summer (DJF) 22.6°C, autumn (MAM) 22.0°C and winter (JJA) 21.2°C (Figure 2.3a-d, respectively).

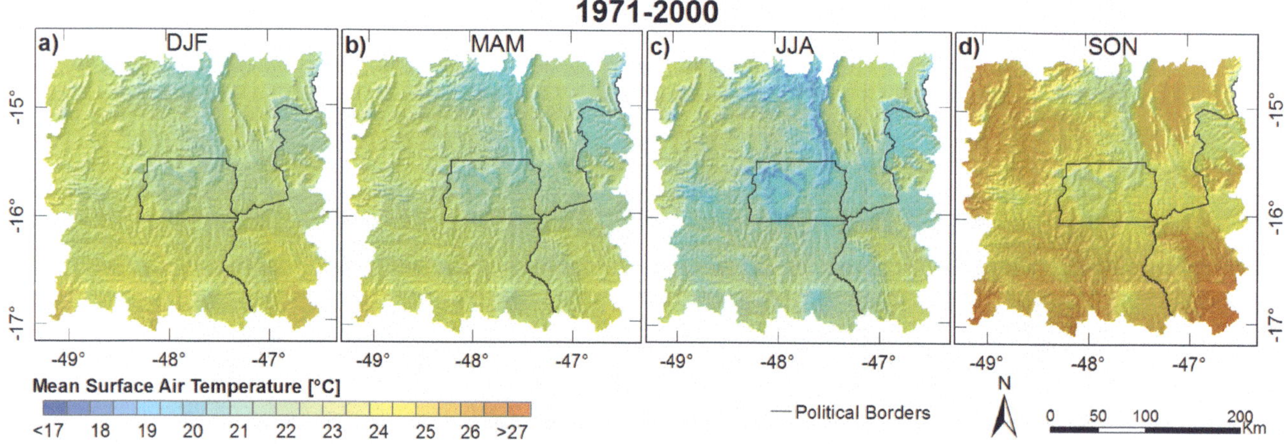

Figure 2.3 Seasonal mean surface air temperature for the DF and surroundings for the normal period 1971–2000, (a) summer – DJF, (b) autumn – MAM, (c) winter – JJA, (d) spring – SON (Borges *et al.* 2014a).

The rainfall amounts for Central Brazil are significantly characterized by seasonality, geographical position and, in part, by topography. The distribution of rainfall over the study area is very heterogeneous for annual precipitation in the north-west and west parts (Figure 2.4a). The average in accumulated annual precipitation for the total study area for the period 1971–2000 is about 1420 mm. However, spatial differences can be more than 500 mm. Borges *et al.* (2014a) identified substantial differences for precipitation in DF, where the western part shows annual precipitations of around 1500 mm, while 50 km to the east, precipitations can be less than 1350 mm. Another typical aspect of the region is the temporal distribution of precipitation during the year, with two very well defined seasons (dry winter and humid summer) (Alves, 2009). In summer (DJF) the average precipitation is 675 mm (Figure 2.4a), while in winter (JJA) the average is 26 mm (Figure 2.4c). Moreover, spring and autumn accumulates 325 and 390 mm of rainfall (Figure 2.4d and 2.4b), respectively.

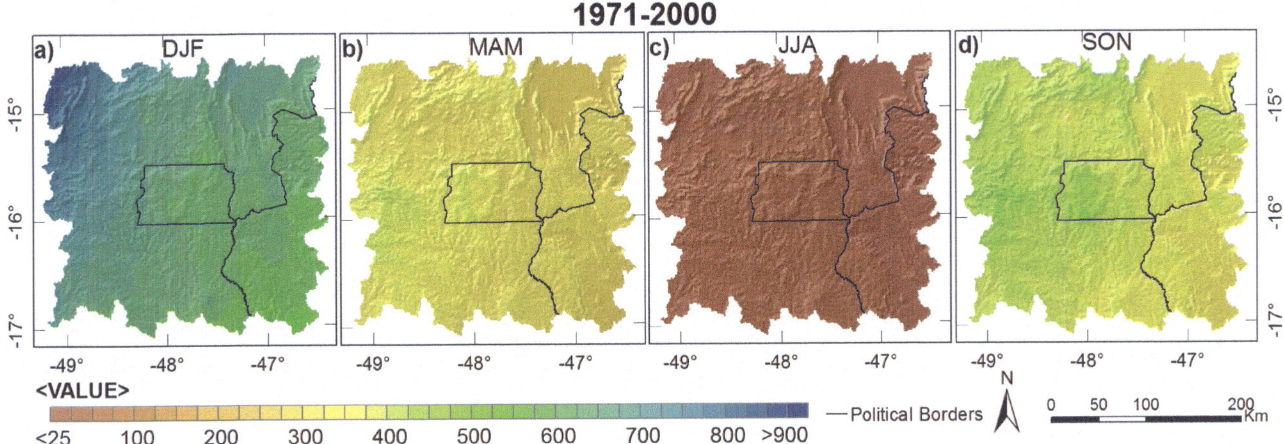

Figure 2.4 Seasonal accumulated precipitation for the DF and surroundings for the normal period 1971–2000, (a) summer – DJF, (b) autumn – MAM, (c) winter – JJA, (d) spring – SON (Borges *et al.* 2014a).

Similar to temperature and precipitation, other climate variables vary significantly over the year. The period from July to September is the driest time and frequently the Civil Defense Forces declares alert state due to the low humidity rates (Jornal de Brasília, 2011; Correio Braziliense, 2012). Figure 2.5a shows the annual variability of relative humidity for the period 1971–2000. Monthly average of 47% of relative humidity is observed in August, while higher numbers are usually observed from October to April. Wind velocity (Figure 2.5b) and global radiation (Figure 2.5c) show similar distribution during the year and are well correlated (r = 0.8). The highest wind velocity and global radiation rates are usually observed from July to September. The annual averages of wind velocity and global radiation for the climatological period 1971–2000 are 2.4 m/s and 1672 j/cm^2, respectively.

2.2.2.3 Changes and trend analysis

A crcucial factor to from future scenarios of climate change is to access the intensity and significance of current climate variations. Therefore, Borges *et al.* (2014a) compare differences between two periods, that is, 2001–2010 and 1971–2010, and perform long-term trend analysis for mean surface air temperature and mean annual precipitation in Central Brazil. Results support the statement that the region experiences already climate warming. On average, differences on annual mean temperature are about +0.7°C. In terms of seasons, spring shows the highest increases in mean temperature of about +1.0°C, followed by summer, +0.7°C. Likewise, the station of Brasília-INMET shows an increase of 0.6°C in annual average (Figure 2.5d). Less severe, the annual maximum surface air temperature increased 0.3°C in the last decade (Figure 2.5e). Long-term trend analysis has also similar results. All stations analyzed demonstrate highly significant trends of annual mean temperature in more than 1.0°C in the last four decades.

Concerning accumulated precipitation, Borges *et al.* (2014a) report alterations in the distribution of rainfall over the year, for instance a decrease of accumulated precipitation in spring (SON) and a more accentuated rainy season (DJF). Variations in mean annual precipitation are mostly negative. Winter shows the highest relative decrease in rainfall, but the absolute amount is negligible. In contrast, spring shows a major rainfall reduction of 12% less than the reference period. On the other hand, summer had an increase in precipitation of about 9%. Figure 2.5f illustrates the disparities between the two climatological periods for the station Brasília-INMET. December and April have the highest increase, while in September and November were observed the most considerable decreases. Correspondingly, the long-term trend analysis performed by Borges *et al.* (2014a) suggest a distinct decrease of precipitation in spring, autumn and, more evident, in winter. For summer, most of the significant trends indicate a positive sign.

Relative humidity, global radiation and wind speed have also faced substantial changes in the last decade. Figure 2.5a confirms the decrease of relative humidity for all months, being more pronounced from July to October. Except for August and September, wind velocity has decreased for all months. October to December was the most affected period. Conversely, global radiation increased almost in all months. The highest increase was observed from August to October.

Lorz *et al.* (2011) demonstrated that land-use changes and water withdrawal in DF are the major causes of decreasing base-flow discharges. However, the observed climate variation may have contributed to the changes in water balance and availability in Central Brazil in the last decades.

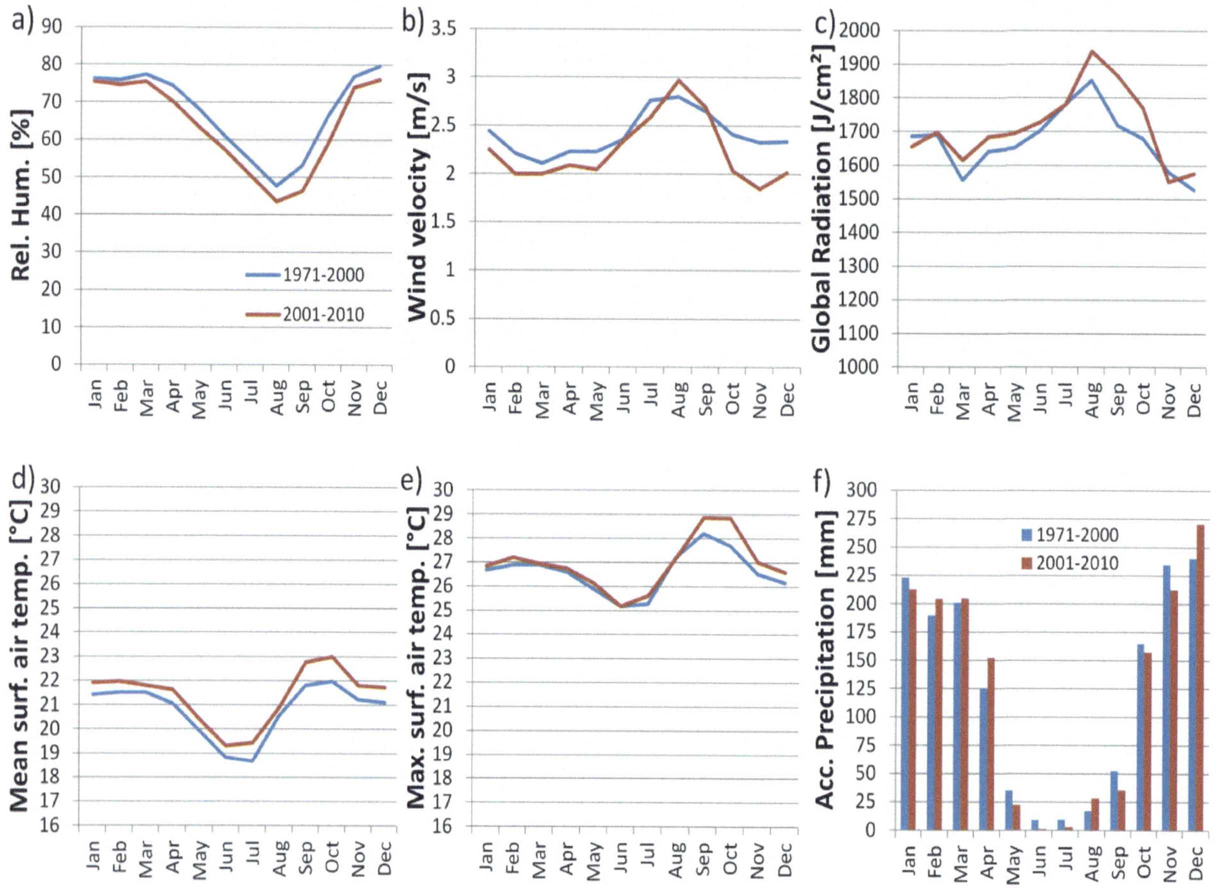

Figure 2.5 Diagram of climate variables for two periods (blue: 1971–2000 and red: 2001–2010). a) mean surface air temperature [°C]; b) maximum surface air temperature [°C]; c) accumulated precipitation [mm]; d) relative humidity [%]; e) wind velocity [m/s]; f) accumulated precipitation [mm] (data source: Brasília-INMET station).

2.3 CLIMATE CHANGE SCENARIOS

Changes in climate conditions and climate variability under the effects of anthropogenic activities, such as emissions of greenhouse gases, can be derived from a variety of data sources using different methods. The most used approach in deriving climate change scenarios is to make use of General Circulation Model (GCM) outputs. However, climate models are not yet capable in representing regional and local physical processes within the climate system (Lu, 2006), and therefore, downscaling techniques are often required in order to obtain information with more detailed spatial resolution (Wilby *et al.* 2004). On the other hand, there are still practical limitations of downscaling methods, especially in case of dubious quality of the meteorological data, insufficient understanding of the links between regional and local climate, and lack of technical capacity (Wilby & Dessai, 2010). Furthermore, high-resolution downscaled scenarios can be misinterpreted as accurate (Dessai *et al.* 2009). Therefore, climate model outputs can be used to derive key information about the changes in future climate without neglecting uncertainties and are often sufficient for impact modeling. To ensure the relevance of the climate scenarios for impact assessment and application, or not, of downscaling techniques, many authors (Lu, 2006; Wilby & Dessai, 2010; Knutti *et al.* 2010) strongly advise users to identify the needs for impact modeling and further assessment. In this topic, the analysis of GCMs outputs by applying the multi-model ensemble method is addressed. Moreover, the possibilities for the development of regional climate scenarios using statistical downscaling approach are demonstrated.

2.3.1 General Circulation Models

Up to now, the Atmosphere-Ocean General Circulation Models – AOGCMs, or GCMs, are the most used tools for simulating the response of the climate system to the anthropogenic activities, such as the increase of greenhouse gases in the atmosphere (Giorgi & Mearns, 2002; Lu, 2006; Marengo, 2007). These models describe the physical and dynamical interactions between the five climate system components: Atmosphere, hydrosphere, cryosphere, continent surface and biosphere (IPCC, 2007). The GCM ability in simulating the climate depends on the region, resolution and climate variable of interest (Meehl *et al.* 2007). Nowadays, the GCM projections are subject of significant uncertainties in the modeling process (Mearns *et al.* 2001), and therefore is very challenging to incorporate these information to impact models, such as hydrologic modeling (Allen & Ingram, 2002). Although most GCMs demonstrate difficulties in producing consistent simulation of precipitation when compared to observations, the temperature simulations are generally well reproduced. Nevertheless, GCMs can deliver useful information about regional climate variations, and, before applying any downscaling technique, the analysis of GCMs outputs is recommended (Christensen *et al.* 2007; van der Linden & Mitchell, 2009).

2.3.1.1 General Circulation Model output: multi-model ensemble and uncertainties

One of the major challenges in climate science is to address the increasing demand of information for impact assessment and adaptation strategies in an environment with (1) expressive differences between the GCM projections; (2) an increasing number of simulations potentially relevant; and (3) an user desire to limit the number of GCMs and simulations (AGO, 2006; Lu, 2006). Although it seems attractive to select climate models based on their capacity to simulate the current climate, there is so far no scientific consensus in a robust method for it (Knutti *et al.* 2010).

Following the approach of multi-model ensemble, recently adopted in several studies on climate change impacts (Giorgi & Mearns, 2002; Hagedorn *et al.* 2005; Nohara *et al.* 2006; Meehl *et al.* 2007; van der Linden & Mitchell, 2009), Borges *et al.* (2014c) investigated trends, consistency and uncertainties of multi-model ensemble for the DF. Variables used are precipitation, temperature and seven extreme indices of all CMIP3 global models. The ensemble projections are also represented as probability density function (PDF) revealing attributes of the data distribution, such as the central tendency and spread, which allow determining probabilistic representation of uncertainty (Wilks 2011). The anomaly is derived from simulations of the 20th century (20C3M) as reference (1961–1990) and future period (2011–2050) based on three scenarios (A2, A1B and B1) of the Special Report on Emissions Scenarios (SRES) (Nakicenovic & Swar, 2000). Figure 2.6 illustrates the median, as well as its trend, spread and level of agreement in the sign of change of the multi-model ensemble for both annual mean surface air temperature and annual precipitation anomalies. Borges et al. (2014c) shows that the warming is evident for all scenarios and seasons when comparing to the 1961-1990 period. An increase of, at least, 0.9 K is suggested until 2050 and the agreement between the simulations in the sign of change becomes more consistent along the decades (Figure 2.6.a). Moreover, in the decade 2041-2050 temperature projections may range from 0.8 to 2.9 K. Precipitation simulations are inconsistent and its ensemble median does not demonstrate significant trend at the annual scale (Figure 2.6.b). A negative precipitation trend of about -10.8 to -15.8 % was detected, in significant levels, only for the dry season (JJA). However, most of the time, the level of agreement is below 80 %, demonstrating therefore the inconsistency between the GCMs in simulating rainfall anomalies for the DF (Borges et al. 2014c).

A negative precipitation trend of about -10.8 to -15.8% was detected, in significant levels, only for the dry season (JJA). However, most of the time, the level of agreement is below 80%, demonstrating therefore the inconsistency between the GCMs in simulating rainfall anomalies for the DF. For extreme indices Borges *et al.* (2014c) expect an increase in heat wave duration index (HWDI), percent of time when minimum temperature >90th percentile value of daily minimum temperature (TN90) and consecutive dry days (CDD). Less significant trends were also identified for intra-annual extreme temperature range (ETR), number of days for which precipitation rate exceeds 10 mm/day (R10) and maximum 5-day precipitation total (R5D). However, the level of agreement in the sign of change is very low. PDFs clearly indicate that changes in precipitation extremes indices are much more uncertain than for temperature indices. The multi-model ensemble combined with percentiles and PDFs can definitely contribute to impact assessment by describing the probability of change in accordance with the level of agreement of simulations. Since risk is defined as the combination of probability of impact and vulnerability (Schanze 2006), the probabilistic approach is of great importance for impact assessment (Giorgi & Mearns, 2002).

2.3.2 Regional climate models

Frequently, the scale of interest for regional impacts requires a much finer resolution from that provided by GCMs (Mearns *et al.* 1999; Lu, 2006). For example, models used to simulate the effects of climate change on water availability usually operate at spatial resolutions varying from micro- (<1500 ha) to macro-scale (>10,000 ha) river basins. Results may be highly susceptible to climate variations that may be hidden in coarse-scale climate variations, particularly complex topography regions, coastlines, and in regions with heterogeneous land surface covers and significant water bodies, such as large rivers and reservoirs (Wilby *et al.* 2004). In the last two decades, a number of techniques have been developed to generate high resolution climate information. Some approaches tend to be very complex and/or computationally extensive, such as dynamic downscaling, while others are simple based and computationally economic, for example, the statistical approaches (Lu, 2006). The adequate technique refers to the needs of the end user and physical factors, such as topography and data availability (Maraun *et al.* 2010). However, the ability to downscale to finer time and space scales does not imply to a higher confidence in the scenarios produced (Wilby & Dessai, 2010).

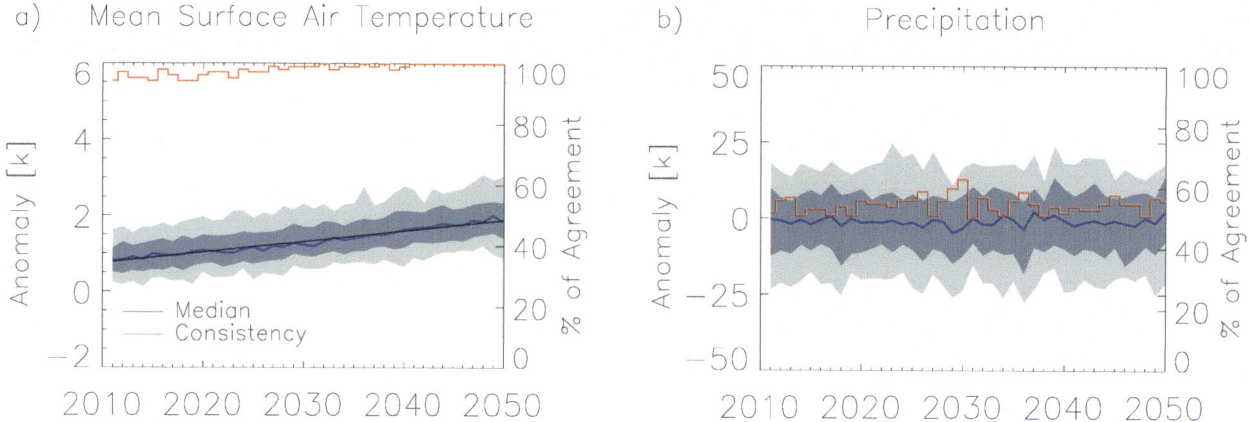

Figure 2.6 Annual anomaly of a) mean surface air temperature and consistency of 103 simulations (22 models × runs × SRES scenarios) and b) precipitation and consistency of 100 simulations (22 models × runs × SRES scenarios). Plots show the median (blue); the consistency (red) according to the % of agreement in the sign of change (right vertical axis); the range between the 10th and 90th percentile (light gray); the interquartile range – IQR (dark gray); and the linear trend of the median (black).

2.3.2.1 Statistical downscaling (SDSM)

The Statistical DownScaling Model (SDSM) is a decision support tool that allows rapid development of local climate change impacts assessments. The tool comprises a robust statistical downscaling technique which facilitates the rapid and low-cost development of multiple single-site scenarios of daily surface climate variables under current and future regional climate forcing (Wilby *et al.* 2002). SDSM is best described as a hybrid of regression-based and stochastic weather generator (Wilby & Wigley, 1997). At the first step, the model identifies the statistical relationship between local variables (predictands) and large-scale circulation indices (predictors) and generates the multiple linear regression equations (Wilby *et al.* 2002). Afterwards, the stochastic element is added to inflate the variance of the multiple linear regression output to better agree with the observed daily data (Diaz-Nieto & Wilby, 2005). Once assumed that this relationship remains constant under climate change, future local climate scenarios can be generated (Wilby & Wigley, 1997; Diaz-Nieto & Wilby, 2005).

The following results show the application of SDSM to the INMET station of Brasília. We considered the mean surface air temperature and precipitation as predictands, while the National Center for Environmental Prediction (NCEP) Reanalysis 1 data was used as predictors for calibration of the model. Once the set of predictors and parameterization of the model are defined, large-scale models are used as boundary conditions for the downscaling of future scenarios.

Before defining the proper set of predictors, the relationship between large-scale predictors and local predictands must be verified by using regression coefficients. In order to avoid a miss choice of the most appropriate NCEP grid cell, data were interpolated to the station of Brasília-INMET using a bilinear method, after Jones (1998). These variables were then screened in order to verify the highest explained variances (R^2) and the lowest standard errors (SE). Sets of predictors were selected also according to their physical relation with predictands. Table 2.1 demonstrates the predictors sets selected in this study. Temperature predictand is related to zonal wind velocity, geopotential height and relative humidity at 500 hPa; geopotential high and relative humidity at 850 hPa; and relative humidity and temperature at surface. Precipitation variance is better explained by zonal velocity and relative humidity both at 500 hPa.

Table 2.1 Selected predictor variables for the station Brasília-INMET.

	p_5u	p500	p850	r500	r850	rhum	Temp
Tmean	X	X	X	X	X	X	X
Prec	X			X			

where p5_u = zonal velocity at 500 hPa, p500 = 500 hPa geopotential height, p850 = 850 hPa geopotential height, r500 = Relative humidity at 500 hPa, r850 = Relative humidity at 850 hPa, rhum = Surface relative humidity, temp = Mean temperature at 2 meters.

Calibration results are summarized in Table 2.2. In this case, it is possible to verify that differences can significantly vary from month to month. Moreover, variances are better explained for mean surface air temperature than for precipitation. As seen in Table 2.2, the determinist part of the model can explain a very low part of the precipitation, and, therefore, SDSM basically becomes a weather generator (Wilby *et al.* 1999).

Table 2.2 Amount of explained variance (R²) and standard error (SE) for each month. In italic the lowest explained variance, in bold the highest explained variance.

		Jan	Feb	Mar	Apr	May	Jun	Jul	Aug	Sep	Oct	Nov	Dec
Tmean	SE	0.88	0.89	0.82	0.79	0.88	0.93	0.93	0.98	1.26	1.12	0.95	0.88
	R²	0.63	0.59	*0.54*	0.57	0.63	0.59	0.67	**0.73**	0.57	0.65	0.62	0.57
Prec	SE	14.8	13.6	14.1	15.7	10.6	7.6	15.7	9.6	10.9	15.0	15.1	13.9
	R²	**0.26**	0.18	0.21	0.21	0.14	0.07	*0.05*	0.13	0.15	0.23	0.18	0.12

Once the calibration is performed and a parameter file is created, the model must be validated. Since the validation procedure must be independent from calibration period (1971–2000), validation is carried out for the period 2001–2010. The proximity between observations and modeled data is a measure of how accurately the calibrated model is likely to downscale any future climate scenario. The comparisons between observed and synthesized climatology are illustrated in Figure 2.7. For mean temperature the downscaled NCEP data replicates the observed climatology rather well, with a bias greater than 0.5°C in June and July (Figure 2.7.a). The scatter plot shows the high level of correlation between observed and synthesized data (Figure 2.7.b). Although less accurate than temperature, the annual distribution of accumulated precipitation is fairly well simulated. Most obvious biases are verified for April, November and December (Figure 2.7.c). The scatter plots confirm the high correlation level between observed and simulated data for the climatological period of 2001–2010 (Figure 2.7.d).

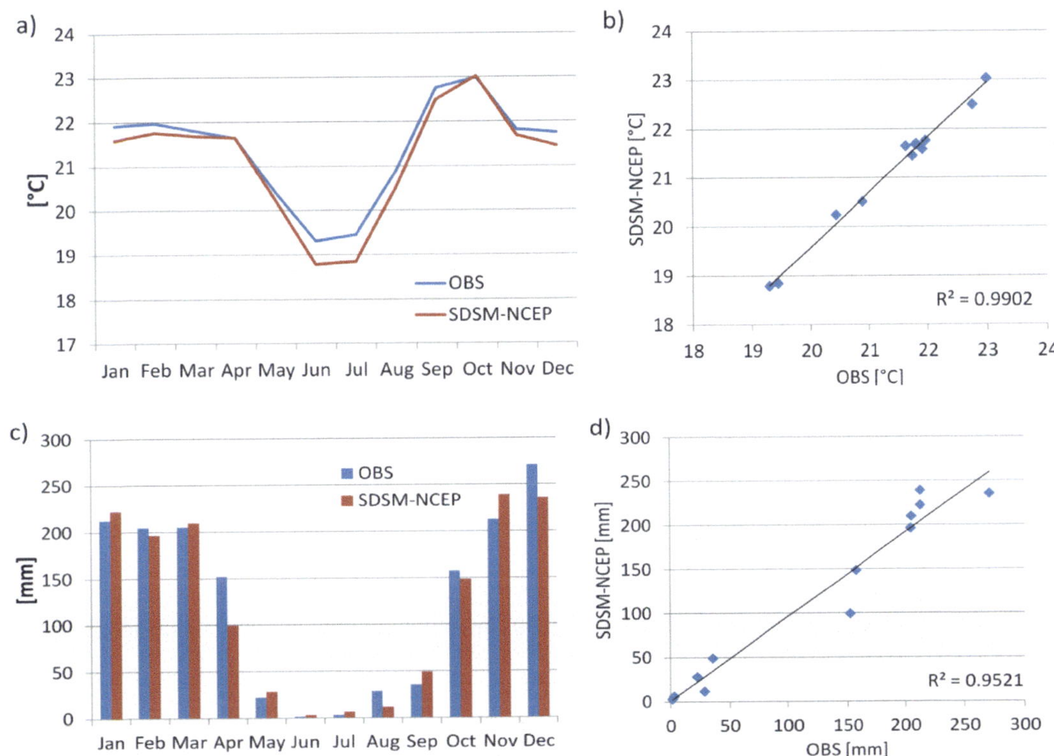

Figure 2.7 Observations vs. synthesized mean surface air temperature (a and b) and precipitation (c and d) for the climatological period of 2001–2010. On left side the monthly values; on the right side the scatter plot.

After validation, the model parameterization is used to perform the downscaling of GCMs predictors, that is, the CGCM3-MR (Flato 2005), ECHAM5 (Roeckner *et al.* 2003) and HadCM3 (Pope *et al.* 2000) models were used. Climatological changes are demonstrated as the comparison of the 30-year mean 2021–2050 against the reference period of 1961–1990 (Figure 2.8a). Monthly projections of temperature climatology show considerable change, especially in September and October. Despite the simulations performed by CGCM3 in January, GCMs here used indicate similar anomalies, and differences among scenarios are very slight. The lowest uncertainty range was obtained in April, with simulations varying from 0.6 to 0.8°C (range of 0.2°C). While January shows the highest uncertainties with simulations oscillating from -0.7 to 0.4°C (range of 1.1°C). ECHAM5 and HadCM3 perform similar anomalies and variation over the year, while CGCM3 simulations are more abrupt varying from -0.7°C in January to 1.7°C in October. A probable reason for that is the lack of capacity of CGCM3 in simulating one or more, here selected, predictors for the region in concern according to the NCEP-Reanalysis data, as demonstrate by Gleckler *et al.* (2008). Moreover, it is noticeable that choice of a GCM as boundary conditions for the downscaling process plays a larger role than the choice of emissions scenario, as previously suggested by Graham *et al.* (2007).

Figure 2.8 30-years average anomaly of mean temperature [°C] (a) and precipitation [%] (b) between the baseline period 1961–1990 and the future scenarios for the period 2021–2050. In red colors the CGCM3 simulation; in green colors the ECHAM5 simulations; in blue colors the HadCM3 simulations.

Figure 2.8b shows that an increase in 30-years mean precipitation is more evident and consistent in January and February. For these months, all models and scenarios agree to an increase in precipitation varying from 4 to 12% and 2 to 16%, respectively. In March, April, November and December an increase is simulated by most of the models and scenarios. While in May, June, July, August, September and October changes are inconsistent, in terms of sign of change, between the models. Likewise to temperature, the lowest uncertainty among simulations is observed in January, deviation from +4 to

+12% (range of 8%); whereas the highest uncertainty was obtained in April, -25 to +27% (range of 52%), followed by August, -35 to 16% (range of 51%). Different from temperature, downscaled precipitation is dependent on both the GCM model and scenario selected as boundary conditions.

The assessment of linear trends of the multi-model ensemble median shows significant and positive trends for mean temperature in all analyzed seasons (Figure 2.9a–e). Annual analysis presents a trend of 0.7°C for the period of 2011–2050. The highest increase obtained was in spring (SON) of about 1°C. Apart from summer (DJF), the level of agreement of the simulations in the sign of change becomes more evident in the last decade. Although the significant trend of the multi-model median in summer, simulations obtained from the CGCM3 model do not agree in the sign of change with the ECHAM5 and HadCM3 simulations. Concerning to precipitation, significant positive trends of the multi-model ensemble median are observed in annual, spring (SON) and summer (DJF) seasons (Figures 2.9f, 2.9i and 2.9j, respectively). It is to notice that CGCM3 simulates the highest precipitation increase in the annual temporal scale, as well as in spring and summer seasons. However, likewise the raw GCM analysis (Borges *et al.* 2014c), the range of uncertainties is very high and no clear agreement among the simulations, not even in the sign of change, is detected on either annual or seasonal temporal scales.

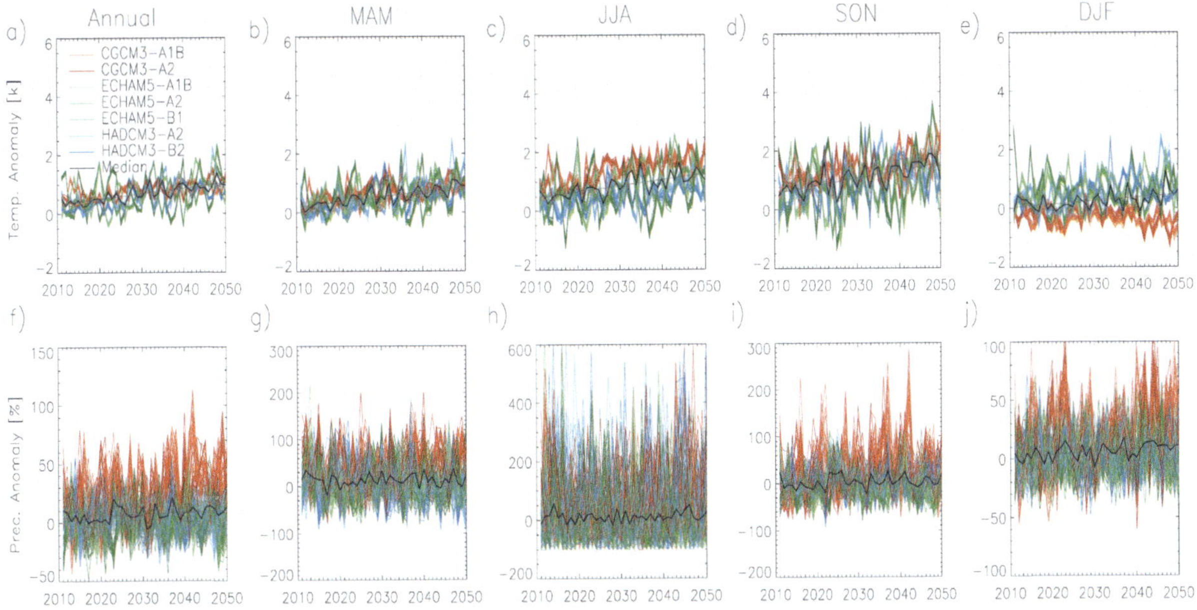

Figure 2.9 Surface air temperature and precipitation anomaly of multi-model ensemble (20-members downscaling simulations x GCMs x SRES scenarios) from 2011 to 2050. Multiple plots illustrate results by climate variables (rows) and respective temporal scale of analysis (columns). Plots show the median (black) of the multi-model ensemble and the linear trend (gray) while significant.

The presented approach, validated for the past, is only applicable in case that the predictand/predictors relationship is assumed to be time-invariant (Wilby & Wigley, 1997; Diaz-Nieto & Wilby, 2005). Potential users are urged to use caution when applying downscaled data for impact assessment, particularly for precipitation. Impact modelers should analyze each site individually, include as much as possible GCMs and validate the model parameterization according to climatic indices of interest (Knutti *et al.* 2010; Maraun *et al.* 2010), for example, variance, percentile 95%, maximum range and percentage of wet days. In addition, a probabilistic climate scenario approach may support the quantification of potential risk of climate change impacts to water resources (New *et al.* 2007).

2.4 REFERENCES

AGO (2006). Climate change impacts & risk management a guide for business and government. Australian Greenh. Off. in the Dept. of the Environ. and Herit., Canberra.

Alexandersson H. (1986). A homogeneity test to precipitation data. *Int. J. Climatol.*, **6**, 661–675.

Allen M. R. and Ingram W. J. (2002). Constraints on future changes in climate and the hydrologic cycle. *Nature*, **419**, 224–232.

Alves L. M. (2009). Clima da Região Centro-Oeste do Brasil. In: Cavalcanti I. F. A., Ferreira N. J., Da Silva M. G. A. J. and Silva Dias, M. A. F. (ed.), Tempo e Clima no Brasil. Oficina de Textos, São Paulo, pp. 235–241.

Ambrizzi T., Rocha T., Marengo J. A., Pisnitchenko A. I., Alves L. and Fernandez J. P. (2007). Cenários regionalizados de clima no Brasil para o século XXI: Projeções de clima usando três modelos regionais. Relatório n. 3, Ministério do Meio Ambiente – MMA, Secretaria de Biodiversidade e Florestas – SBF, Diretoria de conservação da biodiversidade – DCBio. Mudanças Climáticas Globais e Efeitos sobre a Biodiversidade – Sub projeto: Caracterização so clima atual e definição das alterações climáticas para o território brasileiro ao longo do século XXI, Brasília.

ANA (2005). Cadernos de Recursos Hídricos. Disponibilidade e Demanda de Recursos Hídricos no Brasil. Agência Nacional de Águas, Brasília.

Bernhofer C., Goldberg V., Franke J., Häntzschel J., Harmansa S., Pluntke T., Geidel K., Surke M., Prasse H., Freydank E., Hänsel S., Mellentin U. and Küchler W. (2008). Sachsen im Klimawandel, Eine Analyse. Sächsisches Staatsministerium für Umwelt und Landwirtschaft (Hrsg.), Dresden.

Borges P. A., Franke J., Silva F. D. S., Weiss H. and Bernhofer C. (2014a). Differences between two climatological periods (2001–2010 vs. 1971–2000) and trend analysis of temperature and precipitation in Central Brazil. *Theor. and Appl. Climatol.*, **116**, 191–202.

Borges P. A., Franke J., Tanaka M., Weiss H. and Bernhofer C. (2014b). Comparison of spatial interpolation methods for the estimation of precipitation distribution in Distrito Federal, Brazil. Manuscript submitted for publication.

Borges P. A., Barfus K., Weiss H. and Bernhofer C. (2014c). Trend analysis and uncertainties of mean surface air temperature, precipitation and extreme indices in CMIP3 GCMs in Distrito Federal, Brazil. *Environ Earth Sci.*, Advance online publication.

Buishand T. A. (1982). Some methods for testing the homogeneity of rainfall records. *J. Hydrol.*, **58**(1–2), 11–27.

Christensen J. H., Hewitson B., Busuioc A., Chen A., Gao X., Held I., Jones R., Kolli R. K., Kwon W.-T., Laprise R., Magaña Rueda V., Mearns L., Menéndez C. G., Räisänen J., Rinke A., Sarr A. and Whetton P. (2007). Regional Climate Projections. In: Solomon S., Qin D., Manning M., Chen Z., Marquis M., Averyt K. B., Tignor M. and Miller H. L. (ed.), Climate Change 2007: The Physical Basis. Contribution of Working Group I to the Fourth Assessment Report of the Intergovernmental Panel on Climate Change. Cambridge University Press, Cambridge, pp. 847–940.

Correio Braziliense (2012). Distrito Federal entra em situação de atenção devido à baixa umidade. Correio Braziliense 21/03/2013. http://www.correiobraziliense.com.br/app/noticia/cidades/2012/07/27/interna_cidadesdf,313992/ distrito-federal-entra-em-situacao-de-atencao-devido-a-baixa-umidade.shtml, (accessed 21 March 2013).

Craddock J. M. (1956). The representation of the annual temperature variation over central and northern Europe by a two-term harmonic form. *Q.J.R. Meteorol. Soc.*, **82**, 275–288.

Dahmen E. R. and Hall M. J. (1990). Screening of Hydrological Data: Tests for Stationarity and Relative Consistency. Publication No. 49, International Institute for Land Reclamation and Improvement (ILRI), Wageningen.

Dessai S., Hulme M., Lempert R. and Pielke R. (2009). Climate prediction: a limit to adaptation. In: Adger N., Lorenzoni I. and O'Brien K. (eds), Adapting to Climate Change: Thresholds, Values, Governance. Cambridge Univ. Press, Cambridge.

Diaz-Nieto J. and Wilby R. L. (2005). A comparison of statistical downscaling and climate change factor methods: impacts on low flows in the River Thames, United Kingdom. *Climatic Change*, **69**, 245–268.

Dixon W. J. (1950). Analysis of extreme values. *Ann. Math. Statist.*, **21**, 488–506.

Flato G. M. (2005). The Third Generation Coupled Global Climate Model (CGCM3) (and included links to the description of the AGCM3 atmospheric model). http://www.cccma.bc.ec.gc.ca/models/cgcm3.shtml.

Giorgi F. and Mearns L. O. (2002). Calculation of average, uncertainty range, and reliability of regional climate changes from AOGCM simulations via the "Reliability Ensemble Average" (REA) method. *J. Climate*, **15**, 1141–1158.

Gleckler P. J., Taylor K. E. and Doutriaux C. (2008). Performance metrics for climate models. *J. Geophys. Res.*, **113**, D06104.

Graham L. P., Andréasson J. and Carlsson B. (2007). Assessing climate change impacts on hydrology from an ensemble of regional climate models, model scales and linking methods - A case study on the Lule River basin. *Climatic Change*, **81**(Supplement), 293–307.

Hagedorn R., Doblas-Reyes F. J. and Palmer T. N. (2005). The rationale behind the success of multi-model ensembles in seasonal forecasting. Part I: Basic concept. *Tellus*, **57A**, 219–233.

Herzog J. and Müller-Westermeier G. (1998). Homogenitätsprüfung und Homogenisierung klimatologischer Messreihen im Deutschen Wetterdienst. *Deutsch Wetterdienst*, Offenbach.

INMET (1999). Manual de observações Meteorológicas. Instituto Nacional de Meteorologia:. Ministério da Agricultura e do Abastecimento. 3rd ed. Brasília.

IPCC (2007). Summary for Policy Makers. Climate Change (2007). The Physical Science Basis. Contribution of Working Group II to the Fourth Assessment Report of the Intergovernmental Panel on Climate Change. Cambridge Univ. Press, Cambridge.

IPCC-TGCIA (1999). Guidelines on the Use of Scenario Data for Climate Impact and Adaptation Assessment. Version 1. Prepared by Carter T. R., M. Hulme and M. Lal, Intergovernmental Panel on Climate Change, Task Group on Scenarios for Climate Impact Assessment, pp. 69.

Jones P. W. (1998). A User's Guide for SCRIP: A Spherical Coordinate Remapping and Interpolation Package, Los Alamos National Laboratory. http://climate.lanl.gov/Software/SCRIP/, (accessed 5 March 2013).

Jornal de Brasília (2011). Defesa Civil decreta estado de alerta por causa da baixa umidade. Jornal de Brasília 08/07/2011. Retrieved from: http://www.jornaldebrasilia.com.br/site/noticia.php?id=352332.

Kendall M. G. (1970) Rank correlation methods. 4th ed., Griffin, London.

Knutti R., Furrer R., Tebaldi C., Cermak J. and Meehl G. A. (2010). Challenges in combining projections from multiple climate models. *J. Climate*, **23**(10), 2739–2758.

Lorz C., Bakker F., Neder K., Roig H. L., Weiss H. and Makeschin F. (2011). Landnutzungswandel und Wasserressourcen im Bundesdistrikt Brasiliens. *Hydrologie und Wasserbewirtschaftung*, **55**(2), 75–87.

Lorz C., Abbt-Braun G., Bakker F., Borges P., Börnick H., Fortes L., Frimmel F., Gaffron A., Hebben N., Höfer R., Makeschin F., Neder K., Roig H. L., Steiniger B., Strauch M., Walde D. H., Weiß H., Worch E. and Wummel J. (2012). Challenges of an integrated water resource management for the DF, Western Central Brazil: climate, land-use and water resources. *Environ Earth Sci.*, **65**(5), 1363–1366.

Lu X. (2006). Guidance on the Development of Regional Climate Scenarios for Application in Climate Change Vulnerability and Adaptation Assessments. Within the Framework of National Communications from Parties not Included in Annex I to the United Nations Framework Convention on Climate Change, National Communications Support Programme, UNDP-UNEP-GEF, New York.

Maia J. M. F. and Baptista G. M. M. (2008). Clima. In: Secretaria de Desenvolvimento Urbano e Meio Ambiente – Seduma. Governo do DF – GDF, Brasília, pp. 101–109.

Mann H. B. (1945). Nonparametric test against trends. *Econometrica*, **13**(3), 245–259.

Maraun D., Wetterhall F., Ireson A. M., Chandler R. E., Kendon E. J., Widmann M., Brienen S., Rust H. W., Suater T., Themeßl M., Venema V. K. C., Chun K. P., Goodess C. M., Jones R. G., Onof C., Vrac M. and Ehiele-Eich I. (2010). Precipitation downscaling under climate change: Recent developments to bridge the gap between dynamical models and the end user. *Rev. Geophys.*, **48**, RG3003.

Marengo J. A. (2007). Mudanças climáticas globais e seus efeitos sobre a biodiversidade: caracterização do clima atual e definição das alterações climáticas para o território brasileiro ao longo do século XXI. 2. Ed. Ministério do Meio Ambiente, v.1., Brasília.

Mearns L. O., Bogardi I., Giorgi F., Matyasovszky I. and Palecki M. (1999). Comparison of climate change scenarios generated from regional climate model experiments and statistical downscaling, *J. Geophys. Res.*, **104**(D6), 6603–6621.

Mearns L., Hulme M., Carter T., Leemans R., Lal M. and Whetton P. (2001). Climate scenario development. In: Houghton J. T., Ding Z., Griggs D., Noguer M., van der Linden P. J., Dai X., Maskell K. and Johnson C. A. (ed.), Climate Change 2001: The Scientific Basis. Contribution of Working Group I to the Third Assessment Report of the Intergovernmental Panel of Climate Change. Cambridge Univ. Press, Cambridge, 739–768.

Meehl G. A., Stocker T. F., Collins W. D., Friedlingstein P., Gaye A. T., Gregory J. M., Kitoh A., Knutti R., Murphy J. M., Noda A., Raper S. C. B., Watterson I. G., Weaver A. J. and Zhao Z.-C. (2007). Global climate projections. In: Solomon S., Qin D., Manning M., Chen Z., Marquis M., Averyt K. B., Tignor M. and Miller H. L. (ed.), Climate Change 2007: The Physical Basis. Contribution of Working Group I to the Fourth Assessment Report of the Intergovernmental Panel on Climate Change. Cambridge Univ. Press, Cambridge, 747–845.

Nakicenovic N. and Swart R. (2000). Special Report on Emissions Scenarios. A Special Report of Working Group III of the Intergovernmental Panel on Climate Change. Cambridge Univ. Press, Cambridge.

Nohara D., Kitoh A., Hosaka M. and Oki T. (2006). Impact of Climate Change on River Discharge Projected by Multimodel Ensemble. *J. Hydrometeor.*, **7**, 1076–1089.

New M., Lopez A., Dessai S. and Wilby R. (2007). Challenges in using probabilistic climate change information for impact assessments: an example from the water sector. *Phil. Trans. R. Soc.*, **A365**, 2117–2131.

Pope V. D., Gallani M. L., Rowntree P. R. and Stratton R. A. (2000) The impact of new physical parametrizations in the Hadley Centre climate model: HadAM3. *Clim. Dyn.*, **16**, 123–146.

Rapp J. (2000). Konzeption, Problematik und Ergebnisse klimatologischer Trendanalysen für Europa und Deutschland. *Deutsch Wetterdienst*, Offenbach.

Ramos A. M., Santos L. A. R. and Fortes L. T. G. (2009). Normais climatológicas do Brasil 1961–1990. Instituto Nacional de Meteorologia – INMET, Brasília.

Roeckner E., Baeum G., Bonaventura L., Brokopf R., Esch M., Giorgetta M., Hagemann S., Kirchner I., Kornblueh L., Manzini E., Rhodin A., Schlese U., Schulzweida U. and Tompkins A. (2003). The Atmospheric General Circulation Model ECHAM5. Part I: Model Description. MPI Report 349, *Max Planck Institute for Meteorology*, Hamburg, Germany.

Schanze J. (2006). Flood Risk Management – A Basic Framework. In: Schanze J., Zeman E. and Marsalek J. (ed.), Flood Risk Management: Hazards, Vulnerability and Mitigation Measures – NATO Science Series, Volume 67. Springer, Dordrecht, 1–20.

Trewin B. (2010). Exposure, instrumentation, and observing practice effects on land temperature measurements. *WIREs Clim. Change*, **1**, 490–506.

van der Linden P. and Mitchell J .F. B. (2009). ENSEMBLES: Climate Change and its Impacts: Summary of research and results from the ENSEMBLES project. Met Office Hadley Centre, Exeter.

Wilby R. L. and Dessai S. (2010). Robust adaptation to climate change. *Weather*, **65**, 180–185.

Wilby R. L., Dawson C. W. and Barrow E. M. (2002). SDSM – a decision support tool for the assessment of regional climate change impacts. *Environ. and Modelling Softw.*, **17**, 145–157.

Wilby R. L., Charles S. P., Zorita E., Timbal B., Whetton P. and Mearns L. O. (2004). Guidelines for Use of Climate Scenarios Developed from Statistical Downscaling Methods. IPCC Task Group on Data and Scenario Support for Impact and Climate Analysis (TGICA), http://ipcc-ddc.cru.uea.ac.uk/guidelines/StatDown_Guide.pdf.

Wilby R. L. and Wigley T. M. L. (1997). Downscaling general circulation model output: a review of methods and limitations. *Prog. in Phys. Geogr.*, **21**, 530–548.

Wilks D. S. (2011). Statistical Methods in the atmospheric sciences, 3rd edn. Academic Press. Burlington.

WMO (1989) World Climate Data Program: Calculation of Monthly and Annual 30-year Standard Normals, WCDP-No.10, WMO-TD/No. 341, Prepared by a meeting of experts, Washington.

Chapter 3

Protection and exploitation of groundwater resources in Western Central Brazil

R. Stollberg, J. E. G. Campos, W. R. Borges, T. D. Gonçalves, A. Gaffron and H. Weiss

3.1 INTRODUCTION

Brazil is listed as one of the richest countries in terms of water resources worldwide as it accounts for approximately 16% of the world's available freshwater resources (Tundisi, 2005). Moreover, along with its neighbouring countries Argentina, Paraguay, and Uruguay, it occupies large territories of the Guaraní Aquifer, one of the biggest groundwater reservoirs of the world. However, although the country is in general rich in water and the average availability of water across the country is high, particular water stress exists in hydrologically sensitive areas and highly urbanized regions. The Amazonian region, for example, holds about 68% of the Brazilian water resources and covers 45% of the territory but only accounts for 7% of the country's population. Instead, 43% of the Brazilian population lives in the Southeast while this region has only 6% of the total water resources available and just covers 11% of the entire territory (Albuquerque Azevedo & Barbosa, 2011). Besides the spatial variability of the resource's availability and demand, there are also differences regarding its utilization. In Brazil, 45% of the national water resources are used in the agricultural sector, especially for irrigation purposes, 27% are used for urban water supply, 18% serve as industrial waters, and 10% are used for energy production and other issues (ANA, 2005b). According to IBGE (2002), 15.6% of the Brazilian households are supplied with groundwater while 77.8% of cities and communities are supplied by public water supply services that provide a mix of surface water and groundwater resources. The remaining 6.6% are provided by other forms of water supply systems. With respect to this, the complementary water supply is mainly applied in large-scaled cities, whereas groundwater constitutes the main water supply in small to medium sized urbanizations, but also in peri-urban areas of Brazilian conurbations. As farming and widespread agriculture is primarily present in rural regions, their demand, mainly for irrigation purposes, is also covered to a large extent by groundwater.

The DF has implemented a complementary supply system for its urban water supply. Approx. 95% of the water supply is currently covered by surface water resources, whereas only 5% of the water demand is covered by groundwater resources (Moraes *et al.* 2008). According to CAESB (2008b), the usage of groundwater from high potential aquifers for complementary public supply increased permanently over the past years. Meanwhile, about 150 groundwater wells are extracting about 325 l/s for water production (CAESB, 2008a). Besides the primarily used complementary system, some individual areas are entirely groundwater-supplied such as the municipality of São Sebastião in the southeastern DF. As the maximum capabilities of the main water reservoirs of Descoberto and Santa Maria for the urban water supply are almost reached, additional resources and new management approaches are highly required. Thus, the utilization of groundwater resources for the regional urban water supply moves more and more into focus and poses new challenges for a sustainable management of the water resource.

Fundamental differences between surface water and groundwater are based on the different physical and chemical environments in which they occur. Among aquifers, there are huge differences regarding the respective geological environment that influences the water storage capacity and the groundwater flow dynamics. In addition, the geological setting is varying spatially which leads to a hydrogeological diversity

Finally, the groundwater is affected by anthropogenic impacts in a quantitative and qualitative manner. Hirata & Conicelli (2012) recently stated that the great importance of the groundwater resource for the social and economic development contrasts with the lack of awareness regarding its way of exploitation and utilization. Therefore, utilization conflicts on water resources in combination with the ongoing population growth and increasing urbanization emphasize the demand for an effective and sustainable integrated water resource management (IWRM). A balanced groundwater production and an integrated groundwater protection are fundamental objectives in order to fulfil given quality standards. Therefore, a characterization of the aquifer system is of relevant importance to understand and develop sustainable groundwater management concepts for the purpose of an IWRM approach.

3.2 HYDROGAPHIC OVERVIEW

Located at the Central Plateau of the Brazilian Highlands and characterized by hilly terrains and mainly constant altitudes between 1000 and 1200 m, the DF of Brazil comprises watersheds of three national hydrological drainage basins.

These include the Tocantins-Araguaia river basin to the north and the São Francisco basin to the east, while the largest area of the DF is assigned to the Paraná drainage basin towards the south and west (Figure 3.1). The Tocantins-Araguaia basin covers an area of more than 800,000 km . Its major tributary is the Tocantins River, which originates in Goiás State, north of the capital Brasília. It is about 2640 km long and runs from south to north. Towards the east, the São Francisco River arises in Minas Gerais in the Serra da Canastra. It has a length of approx. 3160 km and its drainage basin covers an area of about 630,000 km . The Paraná River, constitutes the second largest river in length among all South American Rivers. With a length of 4880 kilometres, its course passes Brazil, Paraguay, and Argentina before it drains into the Atlantic Ocean (Figure 3.1). Since a watershed comprises all surface water and groundwater resources, soils, vegetation types, and anthropogenic activities of a certain area, watershed-based concepts are widely established in the management of quantity and quality of the strategic water resource (ADASA, 2005).

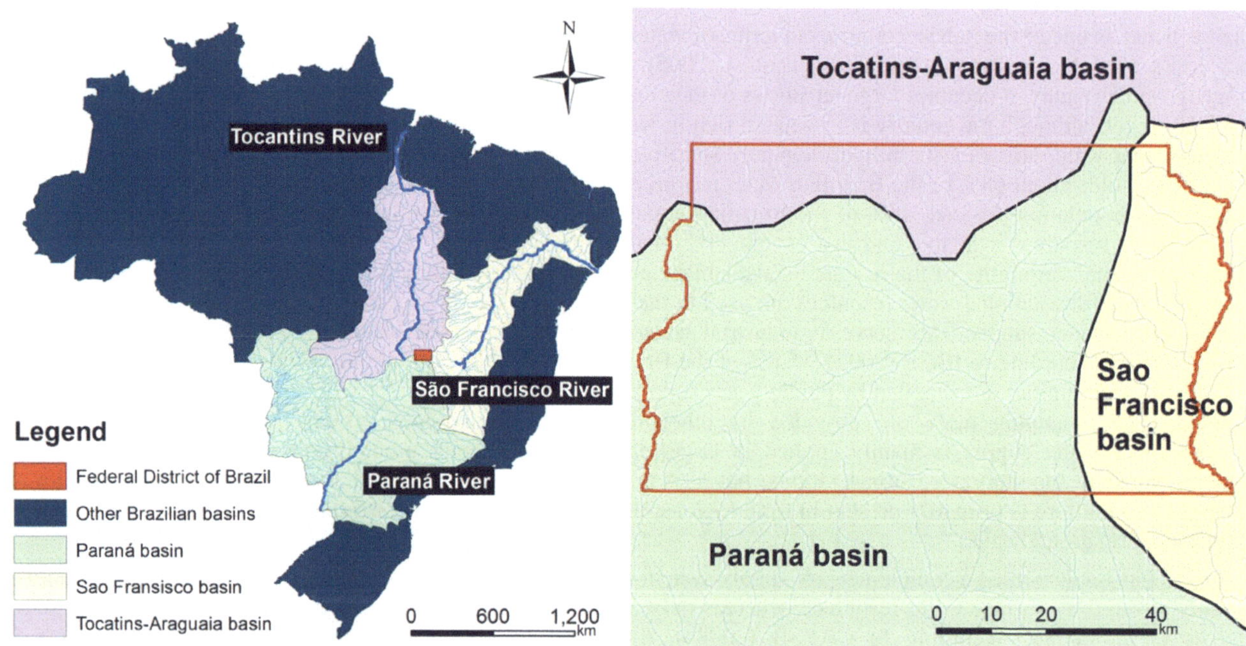

Figure 3.1 National drainage basins and related watershed boundaries in the area of the DF.

The territory of the DF includes three major watersheds (Figure 3.1), while the Paranoá watershed is covering the largest part. As those transition regions of neighboured watersheds tend to be highly sensitive in terms of (ground-) water quantity or probable qualitative interferences, sustainable Integrated Water Resource Management approaches are highly demanded. The DF is divided into seven hydrographic sub-catchments which form the base units of the regional water resource management concept. These units constitute the drainage basins of the rivers Corumbá, Descoberto, Lago Paranoá, São Bartolomeu, São Marcos, Preto, and Maranhão (Figure 3.2). The drainage basins differ in their spatial dimension, hydro(geo)logical setting and water availability. Therefore, it is crucial to understand and characterize the existing hydrogeological system for respective planning, implementation, and optimization of a sustainable water resource management.

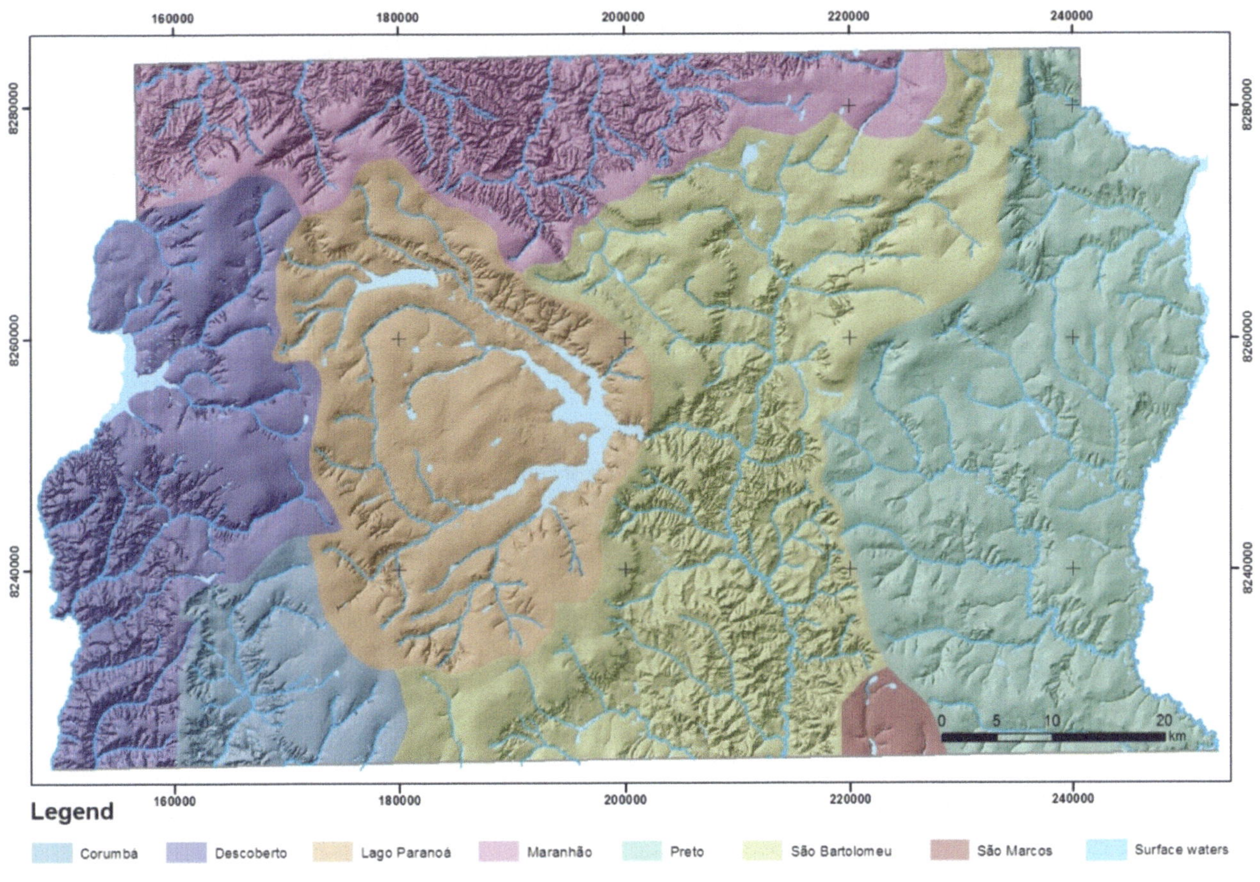

Figure 3.2 Seven hydrographic sub-basins are distinguished for the watershed-based water resource management in the DF.

3.3 HYDROGEOLOGICAL OVERVIEW

The geological framework of the DF is related to the Brasília fold-and-thrust-belt zone which belongs to the major tectonic feature in Central Brazil, the Tocantins Structural Province (Almeida *et al.* 1981). According to Campos (2004), the bedrock formation of the DF is formed by metasedimentary rocks of the Canastra, Paranoá, Araxá, and Bambuí Groups that are of Meso- to Neoproterozoic age (1100-600 mya). The bedrock unit is widely overlain by an extensive lateritic cover of Tertiary and Quaternary age (Mendonça *et al.* 1994). Those lateritic sediments form characteristic soil units such as Inceptisols and Oxisols. Recent colluvium is observed locally but is of minor importance due to its limited spatial occurrence (Table 3.1).

The *Paranoá group* covers about 65% of the entire DF territory and is classified into several sedimentary sub-units from bottom to top, [Q_2], [S], [A], [R_3], [Q_3], [R_4], and [PC] (Figure 3.3). These sub-units are characterized by specified lithology types (Campos, 2004): Sub-unit [Q_2] mainly defines a medium coarse quartzite with intercalated micro-conglomerate bands towards its top. Massive metasiltstones and sandy metarhythmites with locally included lenses of quartzite and micritic metacarbonates form the sub unit [S].

The overlaying unit [A] contains purple, homogeneous, folded slates which comprise sandy interlayers and irregular lenses of quartzite at different levels of the sedimentary sequence. Metarhythmites of sub-unit [R_3] are characterised by an alternating sequence of irregular fine quartzite, bands of metasiltstone, particularly sandy metasiltstone, and metapelites. Siliceous and fractured fine to medium coarse quartzite specify the lithology of unit [Q_3]. Along its spatial extents, clayey metarhythmites of [R_4] are present. These rhythmites comprise bands of quartzite and metapelite of regular thicknesses from one to three centimetres. The following top unit [PPC] contains metacarbonate horizons, lenses and bands of dark quartzite, metasiltstones and metapelites. Rock formations of the *Canastra group* are present on up to 15% of the territory of the DF. These include a high variety on foliated metamorphic rocks such as chlorite phyllite, quartz phyllite, and chlorite-carbonate-phengite phyllite. These phyllite varieties are overlain by interlayers of marble, fine grained quartzite, and cataclasite. The

Araxá group characterizes the smallest lithological formation covering only 5% of the south-western DF territory. This formation is represented by various types of metamorphic schist, including muscovite schist as well as occasional occurrence of chlorite schist, quartz-muscovite schist, garnet schist, and mica quartzite. The different varieties of the *Bambuí group* are mainly observed in the eastern part of the DF and cover about 15% of its area. The *Bambuí group* comprises laminated metasiltstones, clayey metasiltstones as well as banks of metasandstones. These rock formations underwent extensive deformation processes during the *Brasiliano* Orogenesis (Almeida *et al.* 1981, Freitas-Silva & Campos, 1998; Dardenne, 2000).

Table 3.1 Lithostratigraphy of the Distrito Federal, the bedrock setting contains metasedimentary rocks of the Canastra, Paranoá, Araxá and Bambuí group which are of Meso- to Neoproterozoic age (simplified after Freitas-Silva & Campos, 1999).

Age	Group		Description
Neoproterozoic	Bambuí		Essentially pelitic unit composed of schists with rare layers of metaarkoses.
Neoproterozoic	Araxá		Chlorite schists, muscovite-quartz schists, biotite-muscovite schists, rarely garnet-bearing schists. Rarely intercalated fine-grained micaceous quartzites.
Meso-/Neoproterozoic	Paranoá		Psammitic to pelitic unit composed of six distinct units from bottom to top: Q_2 (medium-grained quartzites and intercalated micro-conglomerates) S (metasiltites and metarhytmites with intercalated metacarbonates); A (slates); R_3 (metarhytmites, predominantly psammitic); Q_3 (fine- to medium-grained quartzites); R_4 (metarhythmites, predominantly pelitc) and PPC (metasiltites with lenses of metacarbonates and quartzitic channels).
Meso-/Neoproterozoic	Canastra		Unit composed by sericite phyllites, chlorite phyllites, carbonate phyllites. Quartz-sericite phyllites with intercalated quartzites, metarhythmites and rare lenses of fine-grained marbles near the base of the sequence.

The described lithologies have a significant influence on the present morphology of the DF. In this respect, plateau areas are linked to a great extent to the weathering resistant sub-units [R₃] and [Q₃] of the Paranoá group. In contrast, an increased intermediate dissection level is observed along the slate unit [A] of the Paranoá group and the metasiltstones of the Bambuí group. The morphology along the lithological contacts is intensively rugged due to erosion contrasts. Thus, the geological setting has a significant influence on the altitude profile, valley incision, the surface water pattern, as well as the morphodynamics and landscape evolution of the DF.

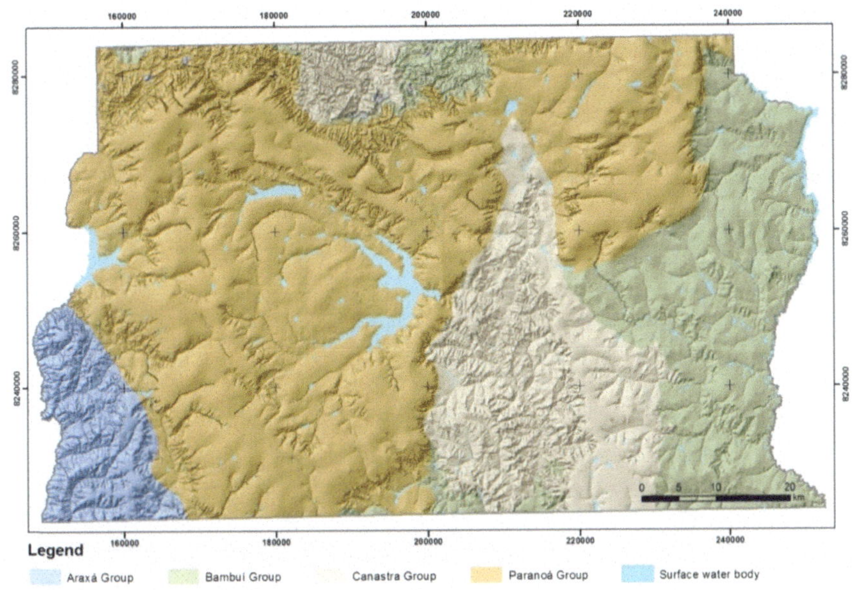

Figure 3.3 The geological setting of the DF is formed by several lithostratigraphic groups. The Paranoá group covers about 65% of the entire territory of the DF.

From a hydrogeological point of view, the subsurface of the DF can be classified into two major aquifer domains: the porous aquifer domain (soils and saprolite) and the fractured aquifer domain (fractured rocks/metamorphic rocks/karstic domain). Since the geological setting involves major heterogeneities regarding the various lithology types, the hydrogeological setting also requires classification into different domains, systems, and sub-systems in relation to its complex characteristics.

The porous domain represents an intergranular system which is interconnected by pore volumes (porosity) and therefore enables the groundwater to move through it. The porous system of the DF emerged from the tropic weathering process of multiple bedrock types and resulted in massive saprolite covers or tropical soils, respectively. The hydrogeological characteristics of the porous medium are mainly controlled by the hydraulic setting, saturated thickness and the respective hydraulic conductivity. Vertical soil thicknesses range from a few centimetres to about 80 meters. Averaged soil thicknesses between 15 to 25 meters are present at about 60% of the area of the DF. The porous domain is predominantly characterized by unconfined aquifer conditions. Up to four different classifications of the porous domain can be distinguished based on the mentioned hydraulic conductivity and vertical thickness properties: P1, P2, P3, and P4. The subsystems P1 to P3 are characterized by vertical thicknesses of >5 meters and the hydraulic conductivity ranges from low to medium and high. Instead, the vertical thickness of P4 is less than 2.5 meters and 1 meter in average, and its hydraulic conductivity is generally low. The porous domain constitutes the transition zone for the infiltrating rainwater and the saturated groundwater water body. Therefore, the porous domain has direct influence on one of the most sensitive parameters of the hydrologic cycle, the groundwater recharge.

The fractured aquifer domain is represented by bedrock formations which are characterized by higher and more constant thicknesses. Instead of pore volumes, these rock units provide cracks, fissures, and fractures of different lengths and orientation for groundwater flow. Hence, the metasediments of the fractured domain need to be considered as heterogeneous and anisotropic aquifer units of limited lateral extension but including high vertical dimensions of several hundred meters, simultaneously. In general, the hydraulic conductivity of the fractured media decreases with depth due to the lithostatic pressure of the overlying formations. The hydrodynamic characteristics of individually fractured aquifer units are generally depending on the rock type while even within the same unit a high variability of permeability is possible. The hydraulic conductivity of the fractured domain is fundamentally controlled by the occurrence of preferential flow paths such as fissures and cracks or fracture density. In general, higher groundwater flow rates are observed within the quartzite formations whereas very low flow dynamics are found in metasiltstones and schist.

The hydrogeological setting of the DF is linked to its geological classification and involves the four aquifer systems: Paranoá, Canastra, Araxá, and Bambuí. According to Figure 3.4, the Paranoá aquifer system is subdivided into the sub-aquifer-systems: [S/A], [A], [R₃/Q₃], [R₄], and [PPC]. The Canastra aquifer system is separated into the sub-aquifers [F] and [Q/F/M]. All fractured aquifers are overlain by the porous aquifer domain that serves as a natural filter on the waters infiltrating from the surface.

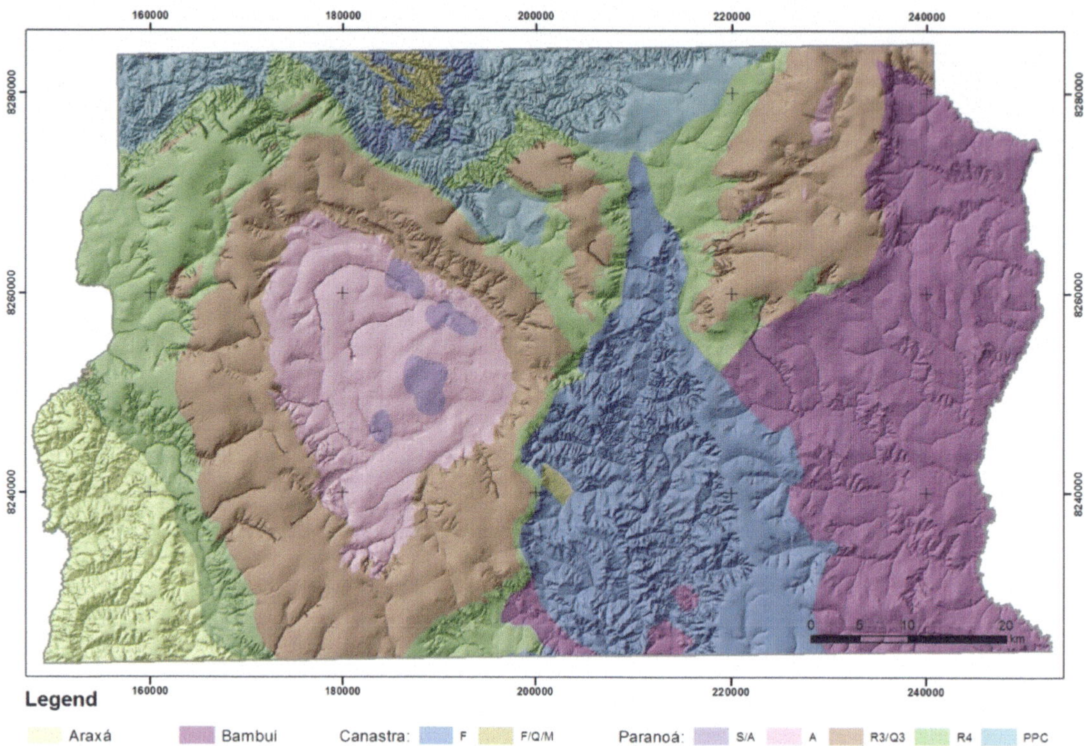

Figure 3.4 The fractured aquifer domain of the DF is classified by several lithological rock types that differ concerning their spatial dimensions and hydraulic properties.

Sustainable groundwater production and a successful implementation of groundwater related IWRM approaches are linked tremendously to a fundamental understanding about the spatio-temporal hydro- dynamics within the hydraulic system.

3.4 CASE STUDIES

As demonstrated previously, quantitative and qualitative designed water management approaches and their successful implementation in terms of an IWRM are fundamentally linked to the understanding and characterization of the subsurface systems in which the water resources occur. In the following, four case studies are presented which were carried out in the DF and are aiming at different research objectives.

(a) Numerical groundwater flow study of the Pipiripau catchment to confirm existing hydrogeological conceptualizations, to verify hydraulic properties derived from field tests and measurements, and to quantify local water balance components
(b) Site investigation of the municipal landfill of Brasília, Lixão do Jóquei, for assessing its contamination potential on surrounding groundwater and surface water resources
(c) Stable isotope mapping for groundwater flow analysis, groundwater age estimation, and vulnerability
(d) Artificial groundwater recharge on tropical soils using treated sewage waters.

While topic (a) involves some general research tasks on hydrodynamic issues on a regional and a local scale; topic (b) addresses a local problem in water quality due to anthropogenic interferences. Case studies (c) and (d) are related to quantitative objectives in terms of aquifer recharge dynamics and innovative water supply.

3.4.1 Numerical groundwater flow modelling for hydraulic system analysis in hydrological sensitive catchments – The Pipiripau river basin

The Pipiripau watershed characterizes a 23,527 hectares large catchment in the northeast of Brasília that constitutes one of the main water supply systems of the DF. Located at the border of the Goiás State, the Pipiripau catchment is used to maintain the water supply of 180,000 inhabitants of the town Planaltina (ANA, 2010). The watershed is intensively used for agricultural production and farming such as grain cultivation, fruit growing, and stock breeding. Agriculturally used areas take up to 71% of the watershed's total area. The environmental impacts by intensive land use are extreme. According to ANA (2010), native vegetation removal, a non-sustainable land use management, and the overexploitation of water resources make up the major environmental problems. The Pipiripau catchment is one of the most critical watersheds of the DF in the scope of a balanced surface water resource management between demand and availability (see chapter 4 for a detailed description of the river basin).

Several water utilization conflicts between various consumers and acceptors have been noticed recently (ANA, 2010). That includes farming, industries, the regional water supply, and sanitation institutions of the DF such as CAESB. Therefore, the government introduced watershed based management strategies to avoid an overexploitation of surface water resources especially during the dry season from June to September. In addition to that, the groundwater resource for additional water supply of the Pipiripau catchment comes to the fore. Meanwhile, about 60 wells are extracting groundwater volumes permanently (ANA, 2010). Several recent studies such as Gonçalves *et al.* (2013), Strauch *et al.* (2013), Camelo (2011), ANA (2010), Albuquerque (2009), and CAESB (2001) point out that an increasing population density as well as an excessive usage of agro-chemicals and irrigation practices will have a limiting effect onto the water resource in terms of water quantity and quality. In this respect, intensive and manifold water resource utilization requires sustainable management approaches.

The objective of the present study is to estimate groundwater flow system related parameters of the Pipiripau watershed by numerical groundwater flow modelling. Recently collected field data was incorporated into this model study to verify and validate the achieved model results. Moreover, non-existent or uncertain groundwater sensitive parameters such as averaged groundwater recharge rates, water flow velocities of the different lithostratigraphic units are investigated in more detail.

The Pipiripau Basin is located at 15°27'14''S and 47°27'47''W (central coordinates). The morphology is ranging between altitudes of 930 to 1206 m a.s.l. Faria (1995) and Campos & Freitas-Silva (1998), characterize the study area by a structural dome with an undulating relief (83%) and slope gradients between zero and eight percent. In respect to climate conditions, the dry season, the dry season from May to September is characterized by high evapotranspiration rates and low values of relative humidity (<15%). In contrast, the rainy season from October to April involves an average precipitation rate of about 1450 mm (INMET, 2009). The average temperature varies from 17°C in June/July to about 22°C between March and September. According to Coimbra (1987), the evapotranspiration varies around 900 mm during the year, and 12% of the total annual precipitation infiltrates the ground and contributes as recharge to the groundwater. Groundwater is simultaneously ensuring and maintaining the minimum flow of rivers during the dry season by its base flow component. The geological setting of the Pipiripau catchment consists of rock varieties of the Paranoá group which are described in detail in section 3.3. The bedrock setting is formed by slate (A), sandy metarhythmite and medium quartzite (R3/Q3) as well as clayey metarhythmite (R4). The fractured bedrock domain is overlain by a porous saprolite/soil unit. The porous domain is represented by oxisols of thicknesses >20 meters and hydraulic conductivities >1.00E-06 m/s (P1 system) as well as inceptisols (P4 system) which are characterized by thicknesses between 2.5 to <1.0 meters and hydraulic conductivities in the range of <1.00E-08 m/s.

As the groundwater flow in fractured and porous aquifers is highly sensitive to respective hydraulic conductivity characteristics, several studies for the estimation of the hydraulic conductivities were carried out to specify the hydraulic systems of the DF in more detail. The characterization of the main aquifer units concerning their hydraulic specifications are essential for assessing hydraulic system related parameters such as groundwater recharge estimation, groundwater flow velocities, water residence times and seasonal hydraulic water level variations. This information forms the basic input of reliable groundwater resource management approaches.

Hydraulic conductivity values used for the numerical groundwater flow simulation are provided by Gonçalves *et al.* (2013). According to Gonçalves (2012), hydraulic conductivity ranges were gathered for respective geological formations (soil formation, porous domain, and bedrock units) by using different infiltration tests such Open-end-hole tests and infiltration tests according to Heitfeld (1979) as well as by reviewing previous studies for example, by Campos and Freitas-Silva (1998) Souza (2001), Lousada and Campos (2005), Almeida *et al.* (2006), and Fiori *et al.* (2010). In addition, hydraulic conductivities of the fractured aquifers were analysed from pumping test recovery data of more than 80 groundwater wells by using different analytical solutions such as Theis (1935), Cooper and Jacob (1946), and Agarwal (1980).

The porous domain, that is, the P1 and P4 systems, is characterized by average hydraulic conductivity values of 1.68E-06 m/s and 3.11E-07 m/s. The highest conductivity values were found in oxisols, while lowest values were estimated for Inceptisol horizons (Gonçalves *et al.* submitted). The average hydraulic conductivity values of the fractured aquifers were estimated as follows: slate (A): 2.06E-06 m/s; sandy metarhythmite/medium quartzite (R3/Q3): 8.43E-07 m/s; clayey metarhythmite (R4): 1.26E-06 m/s. The transition zone between the bedrock formation and the soil horizon is classified as saprolite. According to Gonçalves (2012), the deep weathered bedrock zone shows hydraulic characteristics between those of the soil formation (<1.68E-06 m/s) and the bedrock units (>8.43E-07 m/s).

A numerical groundwater flow model was created to analyse regional groundwater flow dynamics of the Pipiripau catchment by including previously collected structural geological information as well as hydraulic conductivity data. The groundwater flow model is separated into three structural layers. The first layer characterises the soil horizons P1 and P4. The second structural layer reproduces the saprolite unit, while the third layer characterizes the bedrock formation and considers rock materials of the A, R3/Q3 and R4 sub-system. For the structural geological 3-D modelling, lithological information of 62 boreholes was used. The digital elevation model is based on available grid data with a spatial resolution of 20m x 20m. It was derived from the Codeplan (1992) dataset. The finite element model comprises 31,164 model elements per layer and 15,930 computational nodes per slice.

Along the outer model boundary, a second kind no-flow-boundary-condition (Neumann-condition) was assigned along impermeable/less conductive geological contacts. According to Campos and Freitas-Silva (1998), physical hydrological basin boundaries do not necessarily match the hydrologic watershed boundaries but in smaller catchments, such as the Pipiripau basin, water dynamic related differences are marginal. In addition, related surface running waters were characterized by a third kind boundary condition (Cauchy condition). The steady-state flow model was calibrated concerning its hydraulic conductivities and a steady-state mean groundwater recharge. Groundwater recharge was assigned as a top inflow flux using the Neumann boundary condition.

The steady state flow model was calibrated on averaged groundwater levels of 41 observation points reaching a correlation between observed and predicted hydraulic head data of R = 0.996 (Figure 3.5). The calibrated hydraulic conductivities of the respective hydrogeological units are given in Table 3.2. Since hydraulic conductivities from literature were implemented consistently for the P1 and P4 subsystem, the calibrated kf-values of fractured units and the saprolite unit differ locally from the literature values. Based on the results of the steady-state model calibration, conductivity parameters of the saprolite unit, subsystem (A), and subsystem (R4) tend to be generally lower than reported. The differences are in the range of up to one order of magnitude lower. In contrast, the hydraulic characterization of the largest hydrogeological unit R3/Q3 demonstrates a high correlation with the estimated values from literature. These numeric divergences can be explained by high structural uncertainties regarding the hydraulic system since lithological borehole data is limited, spatial information about the fractured subsystems is rare, and literature values are often estimated from local field testing data but are used regionally.

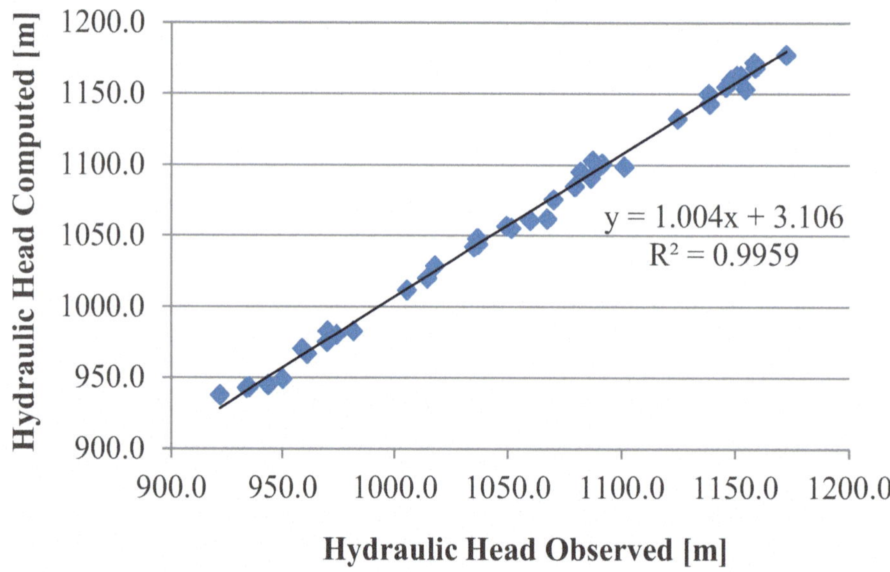

Figure 3.5 Steady-state model calibration results, a correlation coefficient of R² = 0.996 was reached between observed and computed hydraulic head data. Further statistical values are the mean difference Ē = 5.789 and the root mean square error RMSE = 6.972.

Besides hydraulic conductivity, the groundwater recharge was calibrated. Groundwater recharge was spatially assigned at the model surface in model domains which are characterized by terrain slopes less than 2.5%. This GIS-based slope analysis was based on an available digital elevation model with a spatial resolution of 20 × 20 meter. According to Figure 3.5, the best model fit was reached by applying a groundwater recharge rate of 18% of the annual precipitation average, that is, 265 mm/a.

Table 3.2 Overview about calibrated Kf values (hydraulic conductivities) of the calibrated FE-model and respective literature values according to Gonçalves *et al.* (2013).

Hydrogeological Systems	Kf-values (m/s)	Kf$_x$-values$_{model}$ (m/s)	Kf$_y$-values$_{model}$ (m/s)	Kf$_z$-values$_{model}$ (m/s)	Kf-values$_{model}$ (m/s)
Porous Units					
P1	1.68E-06	1.68E-06	1.68E-06	1.68E-06	1.68E-06
P4	3.11E-07	3.11E-07	3.11E-07	3.11E-07	3.11E-07
Saprolite	1.68E-06 - 3.11E-07	5.31E-08	5.22E-08	6.90E-08	2.87E-06
Fractured Units					
A	2.06E-06	1.07E-07	9.80E-08	1.30E-08	7.27E-08
R3/Q3	8.43E-07	3.10E-07	3.32E-07	3.80E-08	2.27E-07
R4	1.26E-06	7.90E-08	4.70E-08	7.00E-09	4.43E-08

The model sensitive range regarding the net inflow was found to be between 17% and 19%. In comparison to groundwater recharge estimates of 12% (Coimbra, 1987) and 33.5% (Carmelo, 2002), the current recharge estimate of about 18% matches the given parameter range from literature and is closer to the recharge estimation given by Coimbra (1987). In addition, the steady-state calibrated net inflow does not include any seasonal variability. Therefore, estimated recharge values tend to be overestimated as precipitation is limited temporally to the rainy season which is characterized by high precipitation rates and high run-off quantities. The surface run-off component does not contribute to regional groundwater recharge. Instead, surface water run-off is collected and drained by surface waters.

The highest groundwater levels of about 1225 m a.s.l. are expected in the northeast and northwest of the catchment (Figure 3.6). The regional hydrodynamics are dominated by a gradient groundwater flow from the recharge areas and plateaus in the west, north, and east towards the centred surface waters. The entire model domain is drained towards the southeast where the lowest groundwater levels are between 925 and 950 m a.s.l. As the regional groundwater flow dynamics are mainly controlled by water inflow (groundwater recharge) at the higher plateau regions, increased groundwater flow velocities are observed along higher slope areas, achieving values of more than 8.0 m/year (Figure 3.7). The model based mean groundwater flow velocity of the Pipiripau catchment is approximately 0.7 m/year. However, modelled groundwater flow velocities are significantly higher in the porous domains (on average 4.6 m/year) than in the lower fractured aquifer domain (1.2 to 0.03 m/year).

Figure 3.6 3-D hydraulic head distribution of the Pipiripau catchment (steady-state); model observation points (flags) and groundwater level measurements.

Figure 3.7 Nodal Darcy flux distribution of the 3-D steady-state flow model; model observation points (flags) and groundwater level measurements.

The presented study focused on the validation of aquifer key properties for an improved groundwater management. Balanced groundwater utilization requires a detailed understanding of the hydrogeological environments and related groundwater occurrences and dynamics. Therefore, the applied modelling study ratifies the pre-existing conceptualization of the DF's hydrogeology. Available hydraulic properties were verified and specified by additional information about groundwater flow dynamics and groundwater recharge. The steady-state model constitutes an initial phase of groundwater related research. Future work needs to include seasonal groundwater variation, baseflow estimation, effects of land-use optimization on groundwater quantity as well as future climate scenarios for supporting a sustainable groundwater management.

3.4.2 Geoelectrical resistivity tomography for groundwater quality characterization at the municipal waste disposal 'Lixão do Jóquei'

The municipal waste disposal 'Lixão do Jóquei' receives domestic waste material of Brasília since the capital's development in the 1960s. With respect to a permanent population increase over the past five decades and a present population of about 2.96 million, the municipal waste disposal site of Brasília handles about 2000 tons of solid domestic waste material every day. These non-recycled materials are dumped superficially at an unsecured area of about 190 hectares western of Brasília's city centre. The dump site does not provide any technical safety measures for groundwater protection such as base liner, drainage system, or wastewater treatment. Thus, the presence of polluted leachate waters constitutes a potential danger for the surrounding environmental compartments and human health since multiple urban settlements, local farming as well as flora and fauna of the Brasília National Park are directly located nearby (Figure 3.8). The presented site investigation focuses on the threat of polluting respective groundwater resources, the spatial characterization of a probable contamination plume, and a water quality assessment.

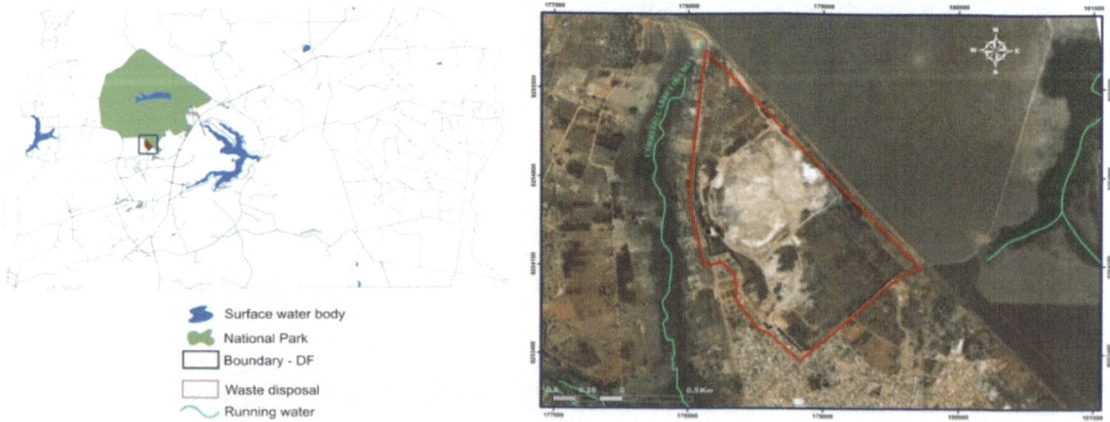

Figure 3.8 Location of the municipal landfill 'Lixão do Jóquei'.

Geoelectrical resistivity tomography was chosen to characterize the structural setup of the underlying hydrogeological system (I) and to spatially characterize probable existing groundwater contaminations based on respective resistivity anomalies (II). To verify the measurements taken during the geoelectrical survey, Direct-Push investigations were carried out for groundwater monitoring well installations and *in situ* groundwater quality analysis (III). The hydrochemical composition of the leachate water was analysed from samples taken at seepage zones from the inside landfill area (IV).

The municipal landfill 'Lixão do Jóquei' (LJB) is located in the Paranoá basin. The surface water system of the Paranoá basin generally drains into the Lago Paranoá reservoir east of the LJB. Three streamwaters exist in close proximity to the LJB site, that is, Córrego do Valo, Córrego do Acampamento, and Córrego Poço d'Água. The origins of the Córrego do Acampamento are based on ground water that drains out along the transition zone between the bedrock and the weathered unit. These spring formations are mainly characterized by the occurrence of very shallow ground waters that are mostly linked to the presence of swampy environments and a significant change in vegetation. In this case, the Buriti palm tree (*Trithrinax brasiliensis*; Neufeldt, 2006) as well as hydromorphic soils and related earthmound fields or 'Murundums' (Oliveira-Filho, 1992) are indicators for shallow groundwater levels. These shallow ground waters as well as nearby surface waters influence the hydraulic situation in the closer surrounding of the LJB. The morphology of the study area is ranging between 1120 and 1130 m a.s.l. Generally, the terrain elevation is dipping slightly from 1250 m a.s.l. in the west to less than 1100 m a.s.l. in the east and southeast. The terrain elevation is dropping along the previously mentioned river valleys in the northwest, west and east of the LJB down to about 1000 m a.s.l.

The geological setting of the study area is primarily formed by rocks of the Paranoá Group that are of Mesoproterozoic to Neoproterozoic age (1100-600 mya). According to Campos (2004), the lower bedrock formation (A) is characterized by bedded, folded and purple coloured schists that can be intercalated by quartzite lenses having foliation features. According to existing (micro-) fractures as well as regional tectonic fracture lines, the bedrock formation is treated as a fractured aquifer unit. The bedrock unit is overlain by a massive and thick saprolite cover that is the product of intensive chemical weathering. This red coloured sedimentary rock is rich in quartz and clay minerals and is still featuring the matrix of the parent rock unit. Due to its granular structure, this Cenozoic sediment is characterized as a porous aquifer.

Holocene fluvial sediments are present in all river valleys. A recent soil cover, consisting of red to yellow Oxisols as well as Inceptisols, forms the top of the geological sequence.

Two-dimensional electrical imaging surveys were applied to retrieve structural information about the hydrogeological system as well as data for the identification and spatial distribution of the present groundwater contamination at the landfill. A multi-electrode array system (SYCAL PRO 72, Iris Instruments) was used for respective data acquisition. The electrodes were installed along a line at the terrain surface using an electrode spacing of 10 meters.

Each section included 1784 nodal resistivity measurements allocated along 30 different depth levels. Such a configuration allows for geoelectrical tomography surveys of up to 70 meters below surface. In total, 15 geoelectrical tomography lines were carried out along the periphery of the LJB (Figure 3.9). Structural interpretation is then based on spatially variable and depth-dependent resistivity data, respectively. Obtained 2D resistivity data was subsequently used to derive structural hydrogeological information as well as information about potential contaminated and non-contaminated zones. Data management, data filtering, as well as topographic data correction were carried out by using the Prosys II software solution (IRIS Instruments). Subsequently, modified resistivity data was analysed by mathematical inversion using the RES2DINV program of the Geotomo Software. For data inversion, least square smoothing (Sasaki, 1989) was applied. Moreover, the Gauss-Newton procedure was chosen for the optimization of recalculated derivatives of the Jacobian matrix during all iterations. The inversion of 2D resistivity data was performed until 3rd or 5th iteration resulting in a root mean square error (RMSE) of 9 to 22%. For the verification of respective spatial interpretations, multiple vertical soundings were carried out by using a mobile Direct-Push technique. Additional information about this technique and further specifications as well as limitations can be found in Schmelzbach *et al.* (2011) and Vienken *et al.* (2012).

Based on these soundings, depth-dependent groundwater samples were taken along the previously surveyed geoelectrical tomography sections by using the SP-16 sampling system of Geoprobe Systems®. In total, 23 groundwater samples were collected from Direct-Push soundings and surrounding surface waters which were directly analysed for pH, temperature, electrical conductivity, and ammonium by using a mobile field device (WTW 350i). In addition to this, a local monitoring network consisting of 14 groundwater monitoring wells was installed along the south-eastern periphery of the landfill (Figure 3.8). Local electrical conductivity and ammonium data were collected to characterize and specify resistivity data obtained from the geoelectrical tomography and to create a geophysical based model of the subsurface setting at LJB.

Two-dimensional resistivity datasets were processed and interpreted to receive structural information about the hydrogeological system as well as to outline potential areas characterized by subsurface groundwater contaminations. Thus, available geochemical information obtained from the local groundwater monitoring/sampling in combination with collected resistivity data was used for an interpreted zoning of uncontaminated areas, areas with low contamination (LC), and areas with high contamination (HC), respectively.

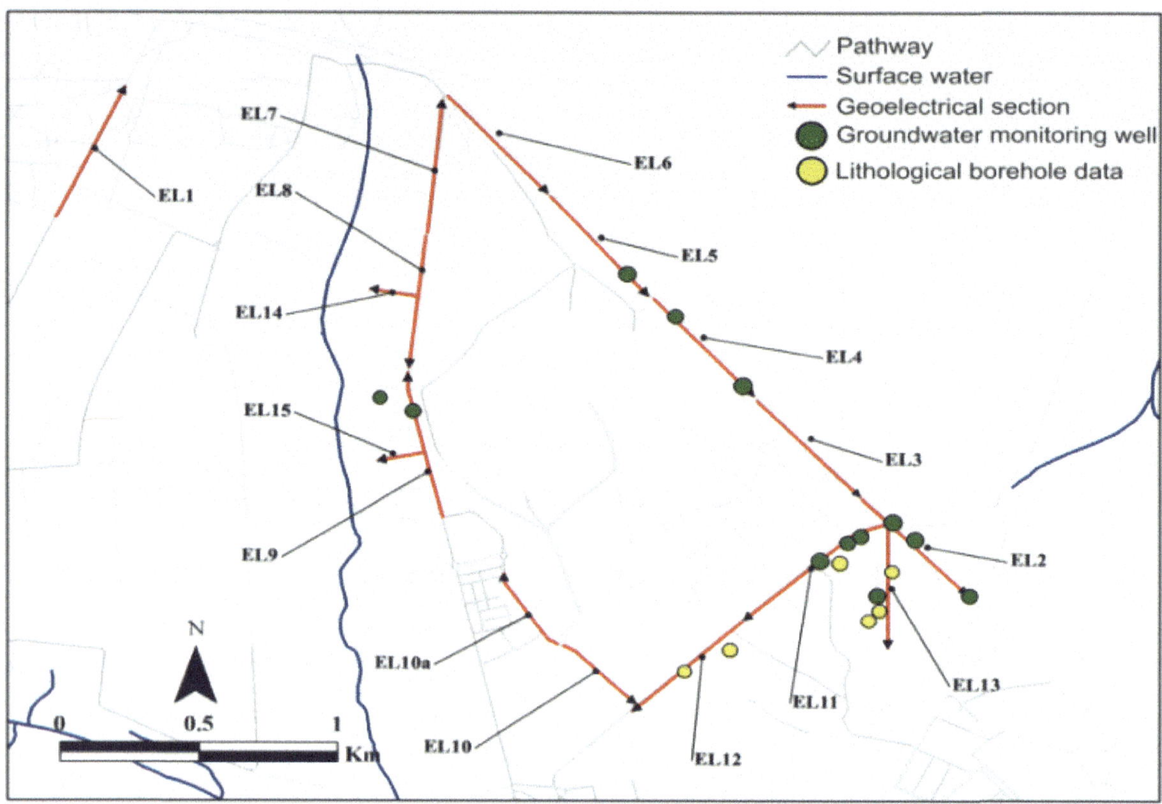

Figure 3.9 In total, 15 geoelectrical tomography surveys (EL1-EL15) were carried out to investigate the structural subsurface setting along the 'Lixão do Jóquei' periphery. Derived structural information is used to create a conceptual hydrogeological model of the landfill.

The zoning is based on a spatial classification matrix including electric conductivity, ammonium concentration, pH-value, and respective resistivity data from geoelectrical tomography surveys. Non-contaminated areas were defined by ammonium concentrations lower than 1 mg/l. These zones were characterized by electric conductivity values below 160 µS/cm and pH values from 6.4 to 7.3. The related electrical resistivity data for non-contaminated areas was higher than 700 ohmmeters (ohm.m) (Table 3.3). For low contaminated zones, resistivity data between 100 and 700 ohm.m was associated with ammonium concentrations between 1 to 10 mg/l, electric conductivity (water) values of 160–375 µS/cm, and pH values from 5.3 to 6.4. Areas with high contamination levels were related to ammonium concentrations larger than 10–30 mg/l, electric conductivities between 375 to 994 µS/cm, and pH values lower than 5.3. These parameter ranges were linked to electrical resistivity values of less than 100 ohm.m.

Table 3.3 Interpreted relations between electrical resistivity values and hydrochemical field data for structural classifications and the indication of contaminated areas at the 'Lixão do Jóquei'.

Zone	Electrical resistivity (ohm.m)	Ammonium NH_4^+ (mg/l)	Electric conductivity in water (µS/cm)	pH
High contamination (HC)	0–100	>10	375–994	5.2–5.3
Low contamination (LC)	100–700	1.0	160–375	5.3–6.4
No contamination (NC)	700–2000	0.0	<160	6.4–7.3

In the following, three geoelectrical sections are chosen for demonstration purposes. Achieved 2D resistivity data could be verified additionally using available lithological borehole information as well as hydrochemical data from local water sampling. With respect to that, one background section without any contamination presence (EL1) and two geophysical lines in contaminated areas (EL11 and EL12) were analysed. The background profile (EL1, uncontaminated) is located approximately 1000 meters northwest of the landfill area (Figure 3.9). The geoelectrical model with an RMSE of 7.8% was obtained after four model iterations and shows electrical resistivity values higher than 800 ohm.m within the entire section (Figure 3.10). Three structural horizons (b) are distinguished on the basis of the received resistivity data (a) (Figure 3.10). Most of the near surface features which are characterized by electrical resistivity values of about 2500 ohm.m and depths > 1150 meter are classified as soils. A second structural horizon shows resistivity data between 800 and 2000 ohm.m, has an averaged thickness of about 10 meters, and is interpreted as the saprolite unit. The lowest structural unit is classified as the bedrock formation that is characterized by electrical resistivities higher than 2500 ohm.m. The groundwater level is expected to be at approximately 5 meters below ground surface.

(a)

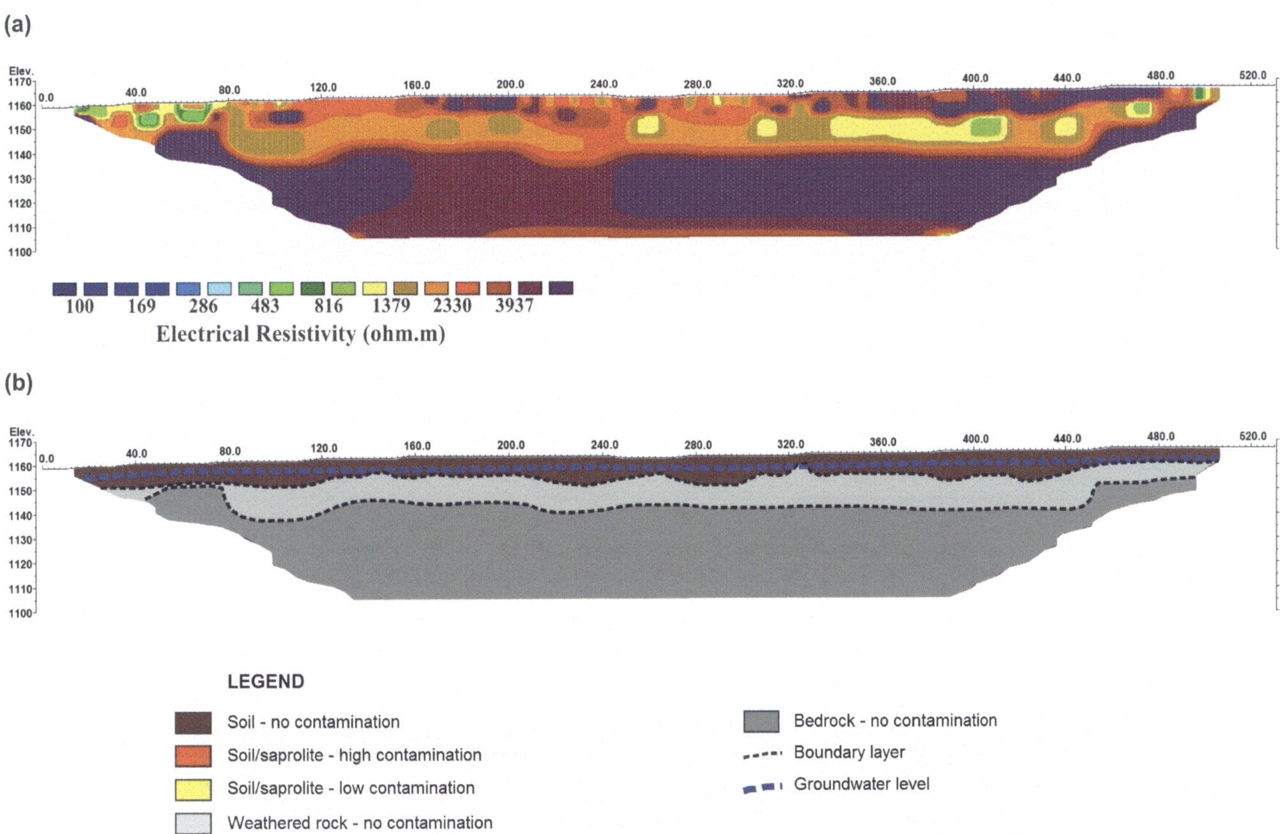

Figure 3.10 (a) Resistivity section of a non-contaminated area; section EL1, (b) Structural model, derived from collected resistivity and geochemical data.

The geoelectrical resistivity survey EL11 was carried out in an area where groundwater contamination was previously proven by Direct Push based groundwater sampling. The resistivity values range from minimum 50 ohm.m to about 4000 ohm.m in maximum (Figure 3.11). This section contains a high conductive zone ranging from the top ground surface down to altitudes between 1045 and 1090 meter. In contrast, a low conductive zone characterizes a resistivity anomaly of more than 3500 ohm.m. The high resistivity zone with altitudes lower than 1045 to 1090 meters specifies the bedrock formation. A high conductive zone, shown in Figure 3.11(a), which is characterized by resistivity values between 50 and 200 ohm.m, is interpreted as clayey soil/saprolite horizons. In the top soil formation, the presence of solid waste material is verified. The illustrated groundwater level is lower than 10 meters below terrain surface and was identified using obtained field data from the locally installed groundwater monitoring network. High resistivity anomalies (200–500 ohm.m) were related to high ammonium concentrations and potentially indicate leachate presence.

In the geoelectrical section EL12, four distinctive resistivity zones can be outlined (Figure 3.12). A near to the surface zone is characterized by electrical resistivities between 300 to 700 ohm.m and is interpreted as an uncontaminated sandy-clayey soil. In contrats, a high conductivity zone with resistivities of about 100 ohm.m represents solid waste material or soil material mixed with solid waste material. The soil unit is underlain by a conductive unit which is interpreted as a low contaminated regolith since the presence of increased ammonium concentrations is proven in this area by local field data. Along altitudes lower than 1095 and 1060 meter, the electrical resistivity increases to more than 2250 ohm.m. This zone characterizes the weathered bedrock or non-weathered bedrock unit.

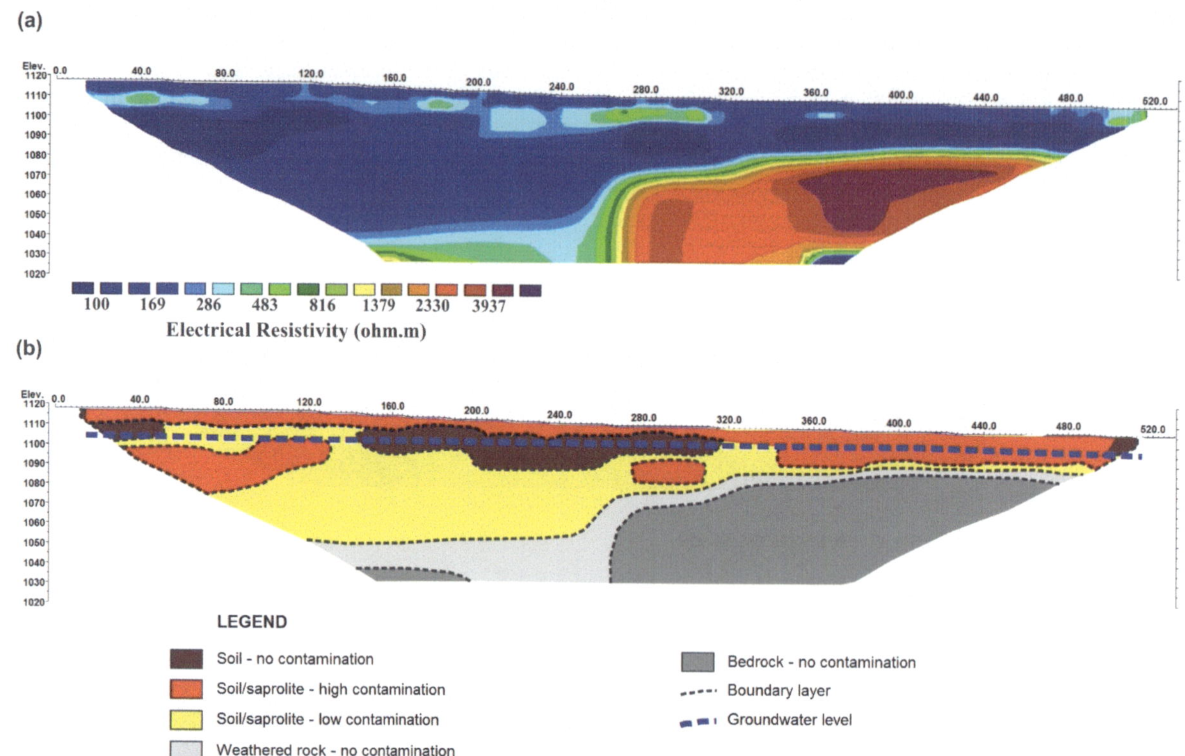

Figure 3.11 (a) Resistivity section of a contaminated area; section EL11, (b) Structural model derived from collected resistivity and geochemical data.

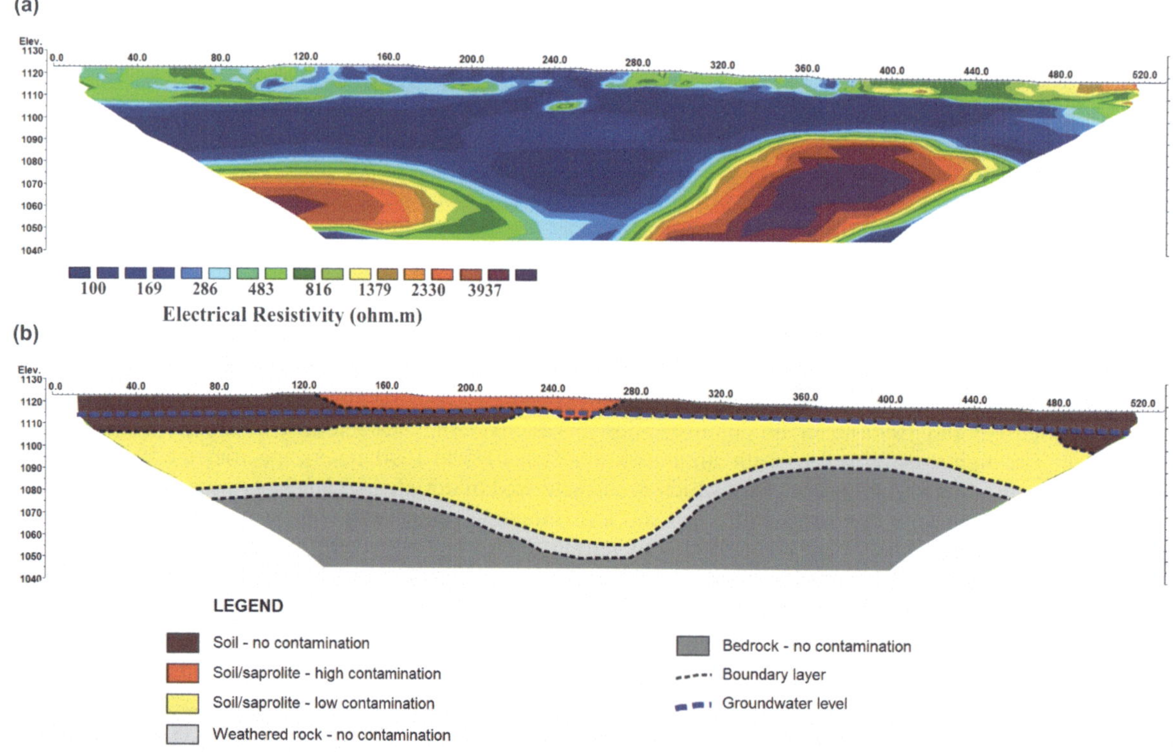

Figure 3.12 (a) Resistivity section of a contaminated area; section EL12, (b) Structural model derived from gained resistivity and geochemical data.

Structural interpretations were carried out on all 2D geoelectrical tomography sections shown in Figure 3.9 and in relation to the hydrochemical parameter classification of Table 3.3.

The results of the spatial interpretation are shown in Figure 3.13. The regional model indicates an increasing presence of probable groundwater contamination zones towards the west and south of the landfill. These zones potentially affect the water quality of the Cabeceira do Valo stream and the urban settlement nearby.

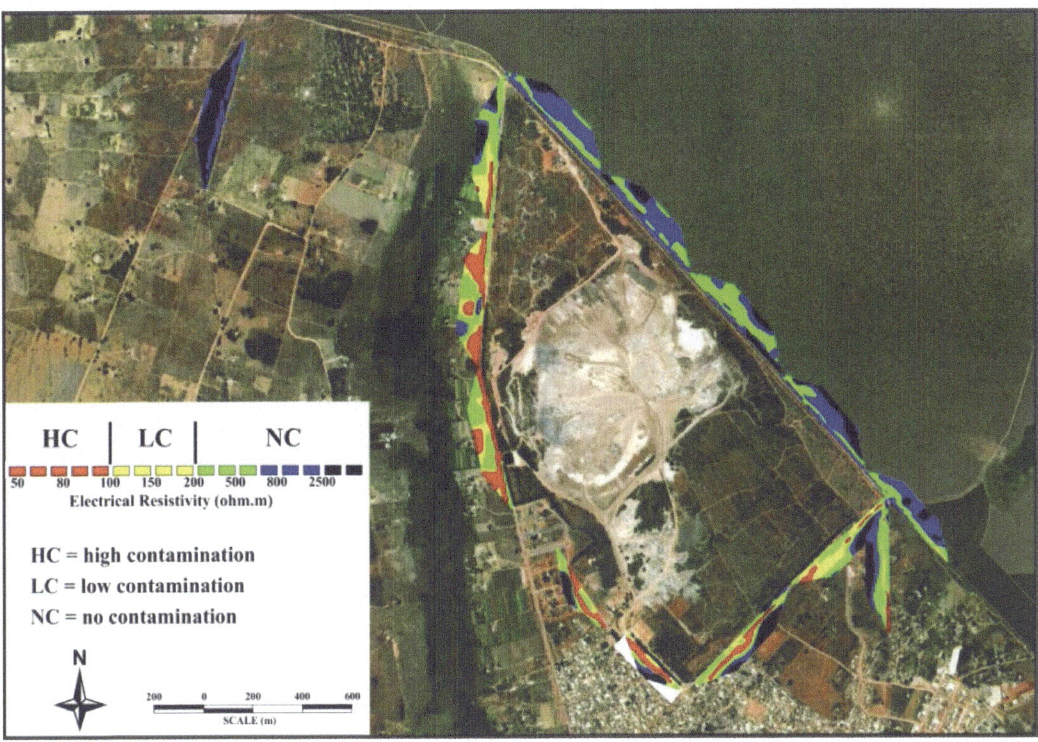

Figure 3.13 Satellite image of the municipal landfill area and the location of respective sections. Analysed profiles indicate an increased groundwater contamination probability in the west, southwest, and south of the landfill 'Lixão do Jóquei' (Cavalcanti, 2013).

For further risk assessment, indicated groundwater contamination zones need to be verified by respective hydrochemical analyses and the findings should be complemented by related mass flux calculations using for example, a numerical modelling approach. However, the resistivity data represents valuable structural information about the hydrogeological setting of the landfill's subsurface. Additional 2D geophysical profiles need to be gained outside of the landfill boundaries to spatially determine the limits of the contamination plumes.

3.4.3 Characterization of groundwater recharge dynamics using stable isotope analyses

The geological and hydrogeological setting of the São Sebastião region is predominantly characterized by the so called São Sebastião Graben, which constitutes a Cretaceous half-graben structure. Within the centre of the graben zone, marble lenses intercalated with phyllites of the Canastra Group were preserved from erosion and form the geological underground of the eastern part of the city São Sebastião. Westwards, metasedimentary rocks such as quarzites, metasiltstones, and slates of the Paranoá Group are present.

With respect to the hydrogeological classification, rock formations of the Paranoá system (R_3/Q_3 and R_4 subsystem) characterise a fractured domain whereas rocks of the Canastra Group represent a fissured-karstic aquifer system. The Canastra System provides a mean yield of approximately 35 m /h of groundwater but single production wells guarantee production capacities of more than 150 m /h. The city of São Sebastião is exclusively supplied with groundwater from local aquifers due to high yields. The local water supplier implemented several water management strategies for sustainable groundwater utilization. This includes the collection and treatment of seepage water, urban waste management, controlled surface water run-off for flood prevention, and maximized groundwater protection areas. With respect to the public water supply that is based on permanently extracted high groundwater volumes, the two main objectives of the case study are:

(a) Where are the recharge zones of the local aquifers and how can the local recharge dynamics of the fractured-karstic aquifer be characterized?

(b) Is there already an overexploitation of the fractured-karstic aquifers happening?

These issues were investigated by using stable isotope mapping and the analysis of groundwater monitoring well data to support a sustainable future water supply.

According to Joko (2002), three possible groundwater recharge mechanisms can be distinguished for the case study region: direct regional recharge, indirect recharge, and local direct recharge (Figure 3.14). Direct local recharge is related to infiltrating rain water, predominantly through structured clayey soils of great thickness. Infiltrating waters reach the saturated zone driven by soil moisture and gravity. The indirect local recharge specifies the water contribution to local surface water streams which receive the aquifer's baseflow coming from highlands. This mechanism is still difficult to quantify since gaining reliable measurements of stream flows is a challenging task due to the various anthropogenic impacts on the local surface waters.

The direct regional recharge specifies rainwater that infiltrates along the highlands which are located towards the west of the São Sebastião urban area. In the highlands, the rain water enters first the massive sandy soils of the porous aquifer. Then, the infiltrating percolate enters the fractures and sub-horizontal fissures of the Paranoá Aquifer System. Due to the high hydraulic gradient, increased flow velocity fields are expected.

Water samples for isotope analyses were collected from existing groundwater wells where lithostratigraphy and well casing specifications are known. For isotope mapping, it was of increased importance that collected water samples entirely originated from the fractured aquifer domain to make sure that no mixing processes of shallow and deep waters took place. Isotopes of hydrogen (^{2}H and ^{3}H) and oxygen (^{18}O) were used to identify local direct recharge and regional direct recharge processes. Therefore, water samples from the high plain area (subsystems R_3/Q_3 and R_4) and the low land area (phyllites and marble lenses) were collected. Results from the isotope analyses were plotted as $\delta(D)$ and $\delta(O)$, see Figure 3.15. Lousada (2005) and Lousada and Campos (2011) already used isotope methods to study existing hydrogeological flow model conceptualisations of the DF presented in Lousada and Campos (2005).

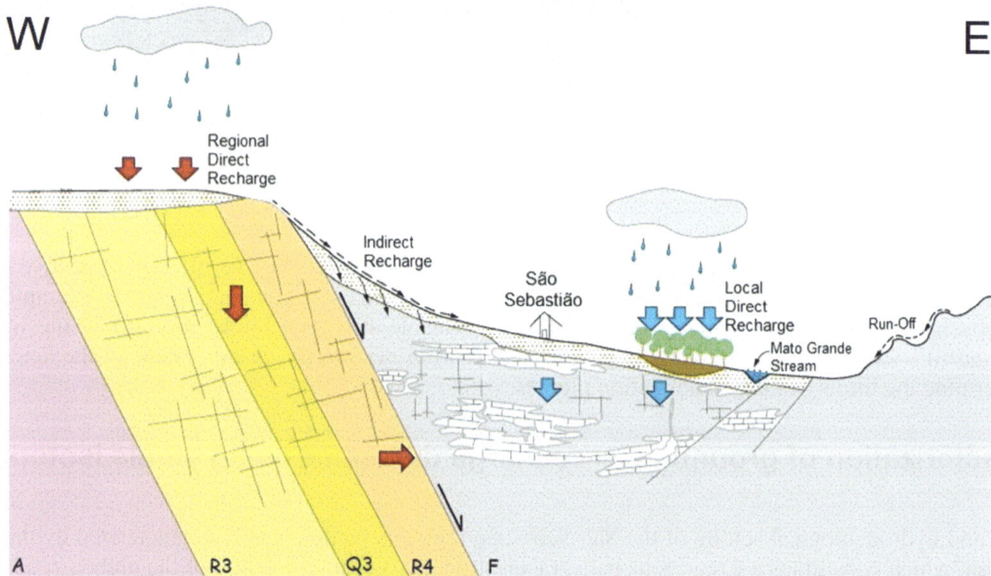

Figure 3.14 The groundwater recharge of the São Sebastião study area is driven by different infiltration mechanisms, schematic cross-section; modified after Mota de Souza (2013).

Figure 3.15 shows the relation of $\delta(D)$ and $\delta(O)$ data in respect to the Global Meteoric Water Line (GMWL) and the local Brasília Meteoric Water Line. The data from groundwater samples show a significant mathematical deviation from those of the meteoric lines. This was to be expected as the infiltrating rainwater had undergone evaporation and evapotranspiration processes which caused isotope fractionation.

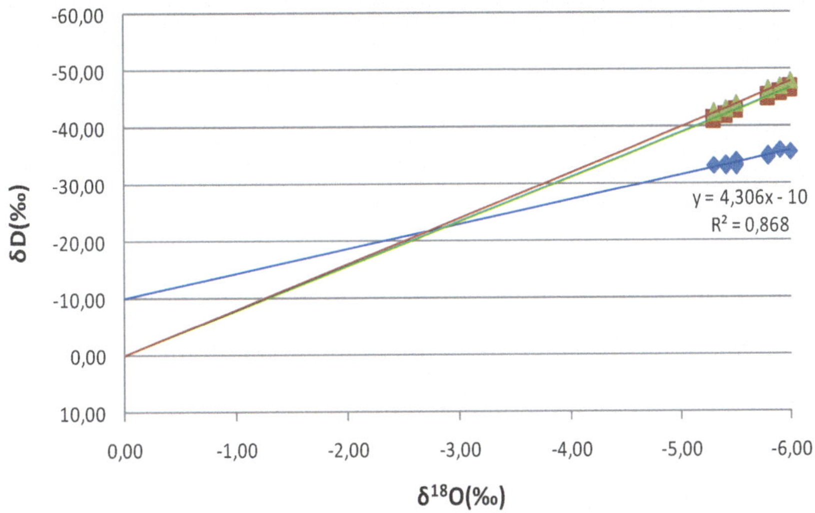

Figure 3.15 Green data line: Global Meteoric Water Line; Red data line: Brasília Meteoric Water Line; Blue data line: δ(D) and δ(O) results of 12 groundwater samples taken from the highlands (4 samples) and lowland areas (8 samples).

According to Pacheco (2011), evapotranspiration effects are mainly observed from water samples that originate from deep water wells of the DF. As the linear function of the δ(D)/δ(O) data of the lowland water samples ($y = 4.4x - 9$) show a much higher mathematical deviation from the GMWL than the one received from the highland water samples ($y = 3.0x - 16.5$), the regional groundwater recharge seems to be the driving recharge mechanism for the groundwater enrichment of the study area. Additionally, local groundwater recharge mechanisms are indicated by increasing levels of nitrate which are observed by local groundwater monitoring (Figure 3.16). The origin of nitrate is to a high extent related to the more intensified agricultural activities in the lowlands. Besides increased nitrate concentrations, high concentrations of sodium and chloride were identified.

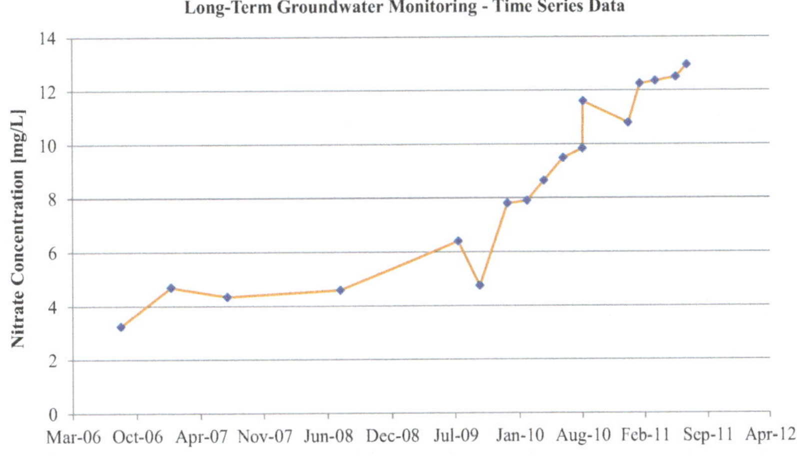

Figure 3.16 Nitrate concentrations (mg/l) from the local groundwater monitoring of the São Sebastião region. Similar effects are observed at additional monitoring wells within the study region.

Tritium data was analysed from 11 groundwater samples that indicated the presence of direct regional recharge. The tritium data (Table 3.4) signify that the groundwater of the plateau area is significantly younger than the groundwater collected in the lowland area of São Sebastião. Lower tritium unit (TU) values or 'older' waters of the lowland area are caused by the isotope fractionation of tritium in groundwater. The higher TU value of the reference location VN-01 is most likely related to the mixing of waters from local and regional recharge processes.

Table 3.4 Tritium data analyses from the deep water wells at the case study site São Sebastião. Tritium units (TU) of waters taken from the highland areas are significantly higher and imply a younger groundwater age than those collected in the lowland areas.

	Well	TU	±2σ (TU)
Highland	MAN-03	1.2	0.2
	SB-03	1.8	0.3
	SB-04	1.4	0.3
	VM-03	1.7	0.3
Lowland (São Sebastião area)	SJ-03	<0.5	—
	SS-01	<0.5	—
	SS-06	<0.5	—
	SS-12	<0.5	—
	SS-13	0.5	0.3
	SS-15	<0.5	—
	VN-01	0.9	0.2

The older waters of the fractured-karstic aquifers predominantly infiltrated in the higher plateau areas and migrated subsequently towards the lowlands or the São Sebastião area by following higher conductive fractures of the R3/Q3 and R4 hydrogeological systems. The topographic gradient, the fractured quartzite aquifer on top of the high plateau area, and increased hydraulic conductivities of the overlying porous domain constitute favourable conditions for rainwater infiltration and regional groundwater recharge.

A risk for the overexploitation of aquifers is given when long-term groundwater extraction exceeds the amount of total groundwater recharge. Conventionally, probable overexploitation is proven through falling groundwater levels in the long term. For the current assessment, data of two groundwater monitoring wells (SS-09 and SS-14) are used to specify potential groundwater storage depletion in the São Sebastião area. These wells are only used to monitor the nearby production wells and are not subject to any groundwater withdrawal. Table 3.5 shows the distances between the monitoring wells SS-09 and SS-14 and the surrounding groundwater production wells. While the monitoring well SS-09 shows a clear influence of the nearby groundwater wells during the annual time series (Figure 3.17), the second monitoring well SS-14 (Figure 3.19) is obviously located outside the area of influence of related production wells since a predominantly invariant groundwater level is observed.

Table 3.5 Distance between groundwater monitoring wells and nearby production wells (CAESB, 2011).

Monitoring well	Production well	Distance (m)
SS-09	SS-08	195
	SS-10	310
	SS-11	385
	SS-07	367
SS-14	SS-13	500
	SS-17	650
	SS-16	720
	SS-02	770

Monitoring data of well SS-09 is indicating various water level fluctuations that are related to long-term groundwater extraction. According to the consistently available time series data for the period from 2006 to 2013 (Figure 3.17) and more detailed water level measurements illustrated in Figure 3.18, a decreasing trend of the water level is clearly indicating an overexploitation of the aquifer. The annual time series data presented in Figure 3.17 show a clear seasonal variability with decreasing water levels from July to October.

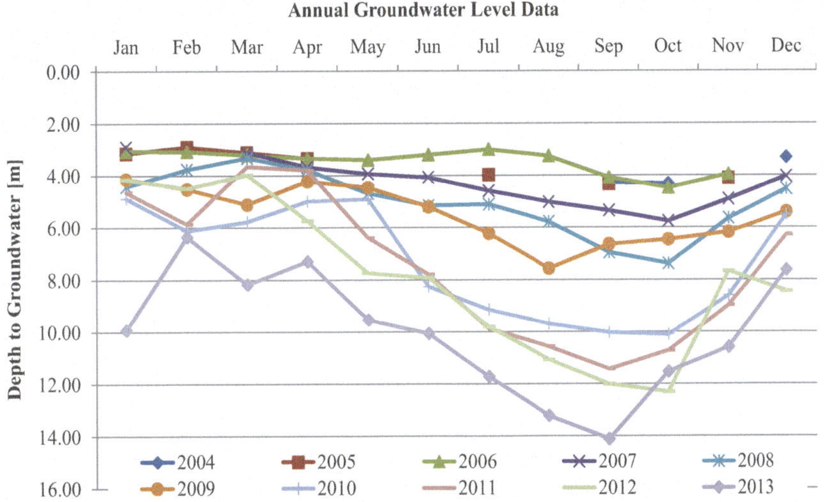

Figure 3.17 Yearly water level time series data of the groundwater monitoring well SS-09 in the São Sebastião study area.

The detailed long-term measurements shown in Figure 3.18 demonstrate even higher contrasts in the water levels between dry and rainy season. The water levels consistently decrease from August to September 2004/2005 to 2013 of approx. 10 m in maximum. Less significant water level contrasts are observed during the maximum of the rainy period in March to April. In the rainy period, water levels were decreasing only about 1.5 meters until 2012. However, a rapid decline is noticed since 2013/2013. Therefore, the data rather implies that groundwater recharge was compensating the water demand during the rainy period in the long term but recently at least locally, the limits of a balanced water system are reached.

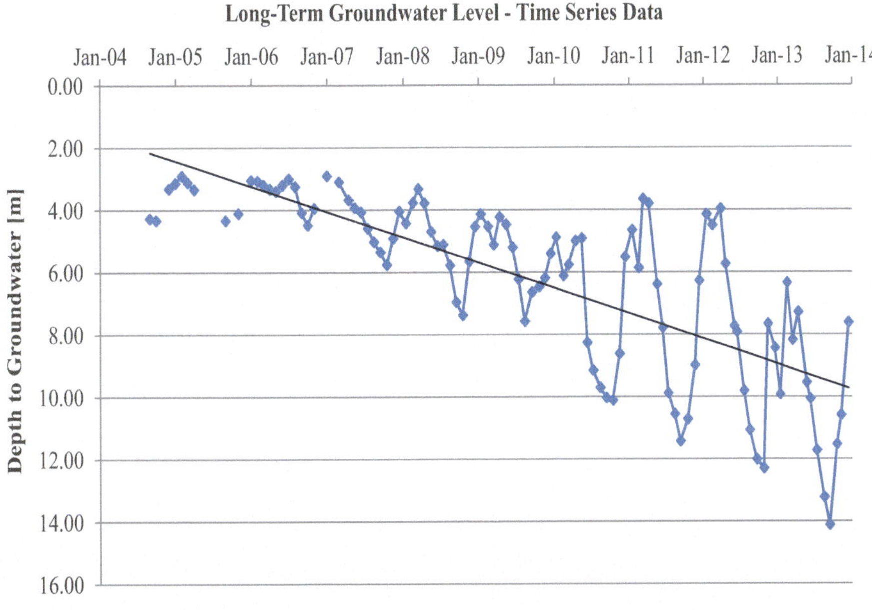

Figure 3.18 Long-term water level variations of the monitoring well SS-09 indicate a decreasing trend.

The annual time series patterns of the groundwater monitoring well SS-14 do not indicate a significant tendency of decreasing water levels (Figure 3.19). Instead, the annual patterns show a consistent behaviour during the years, with seasonal water level fluctuations of about 1 meter between dry and rainy seasons. Detailed long-term time series data presented in Figure 3.20 show lowest water levels during the dry season, whereas even a slight increase of the water level is noticeable at the end of the rainy periods from 2007 to 2012. The analysed water level data of the monitoring well SS-14 do not indicate a general trend of aquifer overexploitation. Further it can be stated that the observed water level represents the regional static water level since the monitoring well SS-14 is located in more than 500 m distance to considered production wells (Table 3.5).

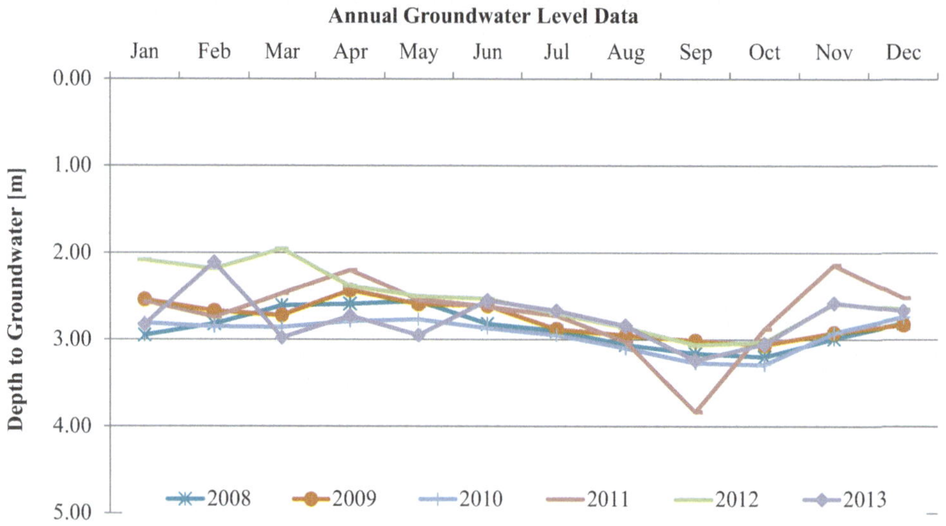

Figure 3.19 Yearly water level time series data of the groundwater monitoring well SS-14 at the São Sebastião study area.

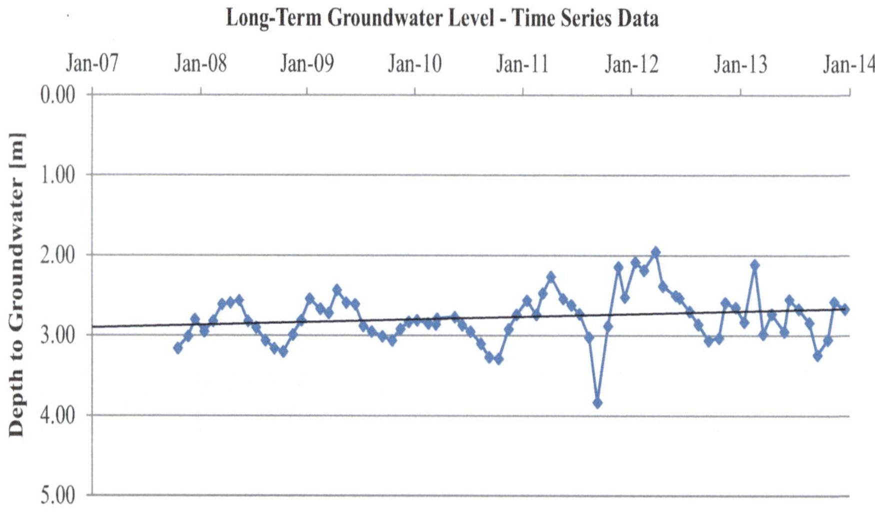

Figure 3.20 Long-term water level variations of the monitoring well SS-14 do not indicate decreasing water levels on the long-term.

Two major recharge mechanisms influence the groundwater recharge dynamics of the aquifers in the São Sebastião study domain, a local direct recharge component which is mainly present in the urban area and surroundings as well as the regional recharge mechanism which is takes place in the nearby high plateau regions. Therefore, it seems to be very important to retain non-sealed areas to enable unhindered rainwater infiltration in both compartments. In the São Sebastião region, it is strongly advised to preserve open space or greened areas as well as urban parks or groundwater protection zones. In relation to the regional recharge, it is highly recommendable to protect high plateau domains from regional condominium development and related soil sealing.

For the assessment of potential groundwater storage depletion, datasets of two local groundwater monitoring wells were analysed. While long-term groundwater monitoring data of well SS-09 clearly shows falling groundwater levels over time and therefore indicates an ongoing aquifer overexploitation, data of the monitoring well SS-14 demonstrate stable hydrological conditions without any signs of overutilization. In terms of annual seasonality, water level data of monitoring well SS-09 denote increasing water level depths due to groundwater abstraction at the end of the annual dry season in August/September. With respect to the annual water level measurements of SS-14, recent groundwater utilisation tends to be balanced since the annual groundwater recharge can compensate water level drawdowns during the dry seasons. In addition, the data even indicate increasing recharge volumes by a slightly rising water level trend.

As the maximum water level drawdowns during dry seasons increase in both monitoring wells, a rising demand on groundwater is clearly demonstrated. Moreover, the annual water level fluctuations between rainy season and dry season also increase in both wells with different intensity. Therefore, the limits of a sustainable groundwater production are approached locally at different levels. In general, further increasing drawdowns during dry seasons will lead to lower dry weather flow rates that automatically will affect water qualities and quantities of the surrounding surface waters. Hence, permanent groundwater monitoring is strongly recommended for revealing regional groundwater storage depletion as soon as possible and to ensure the minimum ecological flow requirements of the hydrological system.

3.4.4 Artificial groundwater recharge for enhanced water supply management in the DF of Brazil

The DF currently has 2.9 million inhabitants and is characterized by a rapid population growth. For 2025, more than 3.2 million inhabitants are predicted (PGIRH, 2006; PGIRH/DF, 2012). The process of urbanization goes along with changes in land use and land cover as well as with an increase of impervious surfaces. Beyond that, the water resources are affected as a consequence of higher consumption and pollution caused by the growing population. The strong impact on the water resources increases further due to an accelerated non-planned urbanization. In addition, the strong seasonality of the climate conditions with its pronounced dry seasons from May to September characterizes the territory of the DF. The existing wastewater treatment plants have reached their limits, especially during heavy storm events. In this context, artificial groundwater recharge might be an efficient method to relieve the pressure on the treatment plants.

Artificial groundwater recharge is a unique technique to take the load off the treatment plants, to store water in the aquifer, and to improve the quality of the water through soil-aquifer treatment (SAT) (Amy & Drewes, 2007; Asano & Levine, 1996; Bouwer, 1999; Bouwer, 2002). The controlled infiltration of pre-treated municipal waste water in the tropical soils of the DF may support the regional groundwater balance and backup high quality water as a sustainable drinking water resource.

The process of controlled percolation (Figure 3.21) of pre-treated sewage ('grey water') comprises a further treatment step. Artificial groundwater recharge has been realized for more than 100 years in Germany and other countries with seasonal variability in precipitation. Artificial groundwater recharge is used to reduce, stop, or even reverse declines of groundwater levels. It allows the storing of water for future and is applied as a wastewater treatment technique (Bouwer, 1978).

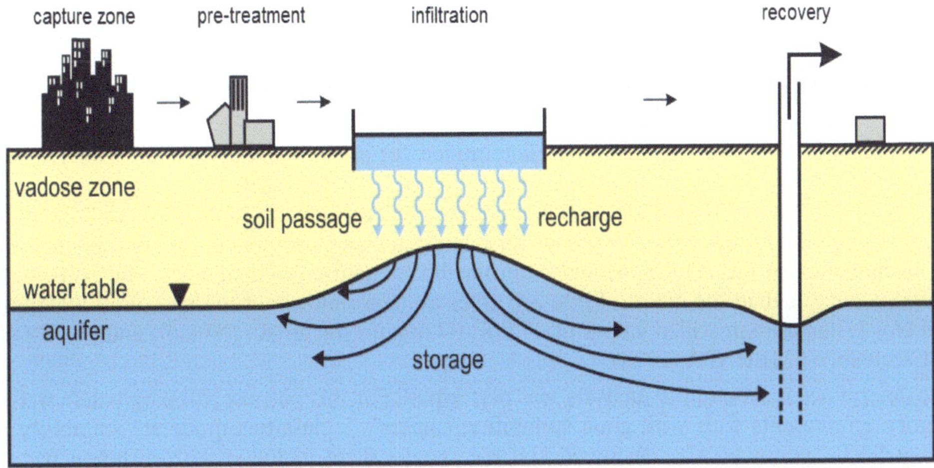

Figure 3.21 System for artificial groundwater recharge using water of impaired quality. The pre-treated wastewater infiltrates through recharge basins into the soil and reaches the water table (modified after Miotliński *et al.* 2010).

The process of artificial groundwater recharge is usually realized in sediments with high hydraulic conductivity. However, sediments or soils with high hydraulic conductivity show a lower potential to retain components of the pre-treated wastewater. The current study focuses infiltration tests based on eight different tropical soils of the DF, Brazil. The focus will be on the identification of suitable soils for a combined treatment and recharge approach and on finding an acceptable equilibrium of hydraulic conductivity and retention potential of the considered soil types.

According to the recommendations of Brazil's Agricultural Research Corporation (EMPRAPA), eight representative soil samples of one meter length were collected for the presented enhanced water supply study. A hydrochemical monitoring on laboratory scale was performed to characterize the sampled soils according to their pollution retention potential. For this purpose, an unsaturated column test was set up. Each soil sample was placed in a cylinder-shaped container of 10 cm in diameter and of 20 cm height. For the infiltration test, wastewater was prepared following German limits for domestic wastewater (DIN 38412 [L24]).

The synthetic wastewater infiltrated from the top, only driven by gravity, to simulate an infiltration process under field conditions. The sampling tank for the percolation water was placed below the soil column. To prevent clay or silt particles from accumulating in the collected percolation water and their possible reaction with the water sample components, a filter system was installed along the outflow. This filter gravel was enclosed by two fine perforated plates and water sampling was done without any contact to the surrounding atmosphere. All test columns were stored in a dark room with stable temperature conditions (approximately 12°C) to prevent the development of biofilms.

In situ-Heitfeld tests were performed to characterize the unsaturated hydraulic conditions (Kf) of the investigated soils. This simple infiltration test is commonly realized in existing boreholes and is applied to soils under unsaturated conditions. In case of a transient implementation of the Heitfeld-test, water is temporally injected until a certain water level increase within the borehole is reached. Subsequently, the natural water level decrease is recorded and analysed (Hiltmann & Stribrny, 1998). All Heitfeld tests were examined under similar experimental setups, such as weather conditions and borehole casing specifications.

The retention potential of the different soil types varies regarding TOC. The inceptisol (soil No. 6) had the highest retention potential for TOC with approx. 99%. Soils No. 2 and No. 3 have also high sorption potentials for TOC (Table 3.6).

Table 3.6 Sorption potentials and measured hydraulic conductivities of tested soils.

	TOC (%)	TN (%)	SO_4^{2-} (%)	PO_4^{3-} (%)	NO_3^- (%)	Kf (m/s)
Soil No. 1	−37.1	−24.4	−22.0	−60.7	−25.5	1.87E-06
Soil No. 2	−86.1	−91.2	24.3	−100.0	−92.7	−
Soil No. 3	−96.3	−89.1	−97.9	−100.0	−82.2	−
Soil No. 4	−55.5	−22.0	3.5	−100.0	−23.2	1.37E-06
Soil No. 5	−36.5	−22.3	−22.0	−100.0	−23.2	3.11E-06
Soil No. 6	−99.0	−98.1	−93.6	−100.0	−95.8	1.12E-08
Soil No. 7	−53.9	−33.2	100.0	−100.0	−76.9	3.01 E-06
Soil No. 8	−50.5	−38.9	9.7	−100.0	7.5	6.21 E-07

Regarding PO_4^{3-}, a retention potential of 100% was determined for all soils except for soil No. 1. The high amount of clay in these soils is responsible for the retention of phosphate (Wild, 1950). Furthermore, the high concentration of iron oxides supports the sorption potential. Regarding nitrate, there are great differences in the observed retention potentials. Soil No. 2 and soil No. 6 showed the best retention potential for this wastewater component. In contrast, low levels of less than 30% nitrate were accumulated in the soils 1, 4, and 5. From the hydraulic point of view, the soils with high hydraulic conductivities are more suitable than the others. Table 3.7 gives an overview about the measured hydraulic conductivities. For soils No. 2 and No. 3, literature provided values of ~1.00E-07 m/s for the alfisol (Nitisol) and 1.00E-06 to 1.00E-07 m/s for the inceptisol (Cambisol) (EMBRAPA, 1978).

To merge all the determined data, a utility analysis was performed. For this purpose, all parameters were categorized into three groups, category 1 represents soils with good suitability, category 2 classifies moderate suitability and category 3 a poor suitability for a SAT-Approach. All considered soil parameters were included with different parameter weightings while the main focus was on the hydraulic conductivity (weighting of 60%) and the retention potential for organic carbon (20%). The retention potential for each of the selected sewage ingredients were weighted with 5%. Table 3.7 displays the derived categories that express the suitability of the tested soil types for an artificial groundwater rechasrge application.

Table 3.7 Final categorization of the soils: category 1 is representing the most suitable soil for an SAT approach.

Soil number	Soil type	Category
Soil No. 7	Red Oxisol	1
Soil No. 1	Alfisol	1
Soil No. 4	Reddish-Yellow	1
Soil No. 5	Red Alfisol	1
Soil No. 3	Inceptisol	2
Soil No. 2	Alfisol	2
Soil No. 8	Reddish-Yellow Oxisol	2
Soil No. 6	Spodosol	2

The categorization results indicate potential soils for a groundwater recharge approach combined with an additional wastewater purification of pre-treated municipal sewage through the soil passage. According to guidelines of EMBRAPA, the Red Oxisol (No. 7), the Alfisol (No. 1), and the Reddish-Yellow Oxisol (No. 4 and No. 8) are the most suitable soils for a soil aquifer treatment. According to the findings of the present study, the soil sample of the Reddish-Yellow Oxisol II (No. 8) was only classified into category 2 with respect to its physicochemical suitability for wastewater purification. It contained a high amount of gravel (~18%) in its topsoil horizon which significantly reduces the potential for purification.

The presented results constitute a fundamental research on an applied SAT approach for characteristic soil types of the DF. For future work, these preliminary results should be further processed in a GIS based approach to identify suitable areas for the soil-aquifer treatment and artificial groundwater recharge.

3.5 CONCLUDING REMARKS

The availability of water resources in the DF is generally limited. Irrigation and water pollution from agricultural activities causes a reduction of available freshwater resources for drinking water supply. In consequence, affected resources are not fully accessible for human water consumption anymore. In addition, polluted water resources such as surface waters or groundwater threaten and influence ecosystems. In consequence, the increasing competition of water utilization requires a regulated and sustainable management approach of the water resources involved. As mentioned before, the water supply of the DF is mainly based on surface water systems. Due to an increasing water demand caused by population growth and an on-going urbanization, groundwater is gaining more importance for water supply and will be irreplaceable in future concepts. However, the development of sustainable best management practices for the groundwater resource requires fundamental knowledge about the aquifer systems in which the waters occur. Information about the hydraulic system such as the structural setting, storage capacity, flow velocity, recharge mechanisms as well as groundwater flow dynamics are highly required to give recommendations for a regulated utilization. Extensive field data about the aquifers was collected during the last decades. Results from the exemplarily implemented numerical modelling study of the Pipiripau basin integrated available datasets and confirmed the existing conceptualization of the DF's hydrogeological system. These model representations can be further processed to create respective groundwater protection and management concepts such as the delineation of groundwater protection zones, vulnerability mapping, groundwater extraction, land-use optimization, future climate scenarios, etc. While these technical measures mainly focus on preliminary actions, further investigations were carried out for assessing regions where water resources are affected by quantitative or qualitative impacts. Hydrologically sensitive regions have to be protected from any qualitative impacts provoked from point pollution sources. Therefore, hazardous locations need to be investigated and assessed concerning their risk potential. Based on the indirect risk to the water resources and human health, appropriate actions need to be developed and implemented. The municipal landfill 'Lixão do Jóquei' represents such a potential point pollution source and was subject of detailed investigation. Geoelectrical tomography and Direct-Push based sounding and sampling were used to identify and localize a probable groundwater contamination at that site. Moreover, areas of limited water resources need to ensure a balanced water production by avoiding overexploitation from aquifers. The water supply of the study region Sao Sebastião is entirely covered by groundwater. Stable isotope mapping was used here to identify the driving recharge mechanisms of the study region. In addition, available monitoring well data was analysed to test for probable overexploitation of the local aquifers. Artificial groundwater recharge by using pre-treated wastewater for infiltration constitutes a promising approach to stimulate available water quantities in seasonal dry periods, droughts, or in case of overexploitation. Therefore, characteristic soil samples were collected and analysed concerning their purification potential for infiltration and recharge. These approaches include enormous transfer potentials and could find application in wide areas of Central Brazil for supporting sustainable groundwater management practices. Although these approaches involve a high potential of transferability, local technical or methodological adjustments will be required for the area of interest. The identification, the development and the process of implementation of any groundwater management approach is fundamentally linked to a consistent monitoring of the focused aquifer system. Therefore, a crucial prerequisite is the collection of long-term piezometric and hydrochemical time series data from a comprehensive aquifer monitoring network to recognize potential hazards and to implement appropriate measures. Groundwater monitoring data can thus provide an essential linkage between policy making and regulatory implementation for the protection and safe exploitation of groundwater resources.

3.6 REFERENCES

ADASA (2005). Caderno Distrital de Recursos Hídricos. Plano de Gerenciamento Integrado de Recursos Hídricos do Distrito Federal – PGIRH. Brasília, DF, p. 111.

Agarwal R. G. (1980). The new method to account for producing time effects when drawdown type curves are used to analyze pressure build-up and other test data. SPE 9289 – 55th SPE Annual Technical Conference and Exhibition, Dallas, TX, Sept. 21–24.

Albuquerque R. W. (2009). Levantamento de malha fundiária e diagnóstico ambiental das propriedades da bacia do Pipiripau (DF/GO) para a participação no Programa Produtor de Água. Dissertation. University of Brasília.

Albuquerque Azevedo H. A. M. and Barbosa R. P. (2011). Gestão de recursos hídricos no distrito federal: uma análise da gestão dos comitês de bacia hidrográficas. *Ateliê Geográfico*, **5**(13), 162–182.

Almeida F. F. M., Hasui Y., Brito Neves B. B. and Fuck R. A. (1981). Brazilian Structural Provinces: an introduction. *Earth Science Review*, **17**, 1–19.

Almeida L., Resende L., Rodrigues A. and Campos J. E. G. (2006). Hidrogeologia do Estado de Goiás (Hydrogeology of the Goiás State). Superintendência de Geologia e Mineração (Superintendency of Mining Geology). Goiás, Brazil. Technical Report, p. 230.

Amy G. and Drewes J. (2007). Soil aquifer treatment (SAT) as a natural and sustainable wastewater reclamation/reuse technology: Fate of wastewater effluent organic matter (EfOM) and trace organic compounds. *Environmental Monitoring and Assessment*, **129**(1), 19–26.

ANA, Agência Nacional de Águas (2005a). Panorama da Qualidade das Águas Subterráneas no Brasil.

ANA, Agência Nacional de Águas (2005b). Disponibilidade e Demandas de Recursos Hídricos no Brasil. Brasília, p. 134.

ANA, Agência Nacional de Águas (2010). Relatório de Diagnóstico Socioambiental da Bacia do Ribeirão Pipiripau ('Environmental Report of Pipiripau basin'). URL: http://www.ana.gov.br (accessed 04 May 2011)

Asano T. and Levine A. D (1996). Wastewater reclamation, recycling and reuse: past, present, and future. *Water Science and Technology*, **33**(10), 1–14.

Bouwer H. (1978). Groundwater Hydrology. McGraw-Hill Publishing Co., New York, p. 492.

Bouwer H. (1999). Artificial recharge of groundwater: systems, design, and management. In: L. W. Mays (ed.), Hydraulic Design Handbook. McGraw-Hill Publishing Co., New York, p. 1024.

Bouwer H. (2002). Artificial recharge of groundwater hydrogeology and engineering. *Hydrogeology Journal*, **10**, 121–142.

CAESB – Environmental Sanitation Company of the DF (2001). Plano de Proteção Ambiental do Ribeirão Pipiripau ('Environmental Report of Pipiripau basin'). URL: www.caesb.df.gov.br (accessed 26 April 2010)

CAESB (2008a). Sinopse dos Poços da CAESB em Áreas Urbanas – Ano 2007: Resumo da situação dos poços da CAESB em áreas urbanas, p. 74.

CAESB (2008b). SIAGUA – Sinopse do Sistema de Abastecimento de Água. Assessoria de Planejamento, Programmação e Controle – PPC. 17 edição, Versão 1.00, Brasília, Destrito Federal, p. 143.

Camelo A. P. S. (2011). Quantificação e valorização do serviço ambiental hidrológico resultante da recomposição de passivos ambientais na bacia hidrográfica do Ribeirão Pipiripau. Dissertation. University of Brasília.

Campos J. E. G. and Freitas-Silva F. H. (1998). Hidrogeologia do Distrito Federal. In: Inventário hidrogeológico e dos recursos hídricos superficiais do Destrito Federal. Parte I. Vol. II. IEMA-SEMATEC/Universidade de Brasília. (Inédito). p. 66.

Campos J. E. G. (2004). Hidrogeologia do Distrito Federal: Bases para a gestão dos recursos hídricos subterrâneos. *Revista Brasileira de Geociências*, **34**(1), 41–48.

Cavalcanti M. M. (2013). Aplicação de métodos geoelétricos no delineamento da pluma de contaminação nos limites do aterro controlado do jokey clube de Brasília. Dissertação de maestrado, Universidade de Brasília, Brasil, p. 128.

Coimbra A. R. S. R. (1987). Balanço hídrico preliminar do Distrito Federal. In: Inventário Hidrogeológico e dos Recursos Hídricos Superficiais do Distrito Federal, Brasília, IEMA/SEMATEC/UnB, (Vol. II, technical report), 50–78.

Codeplan (1992). Mapas topográficos plani-altimétricos digitais do Distrito Federal na escala de 1:10.000, Brasília.

Cooper H. H. and Jacob C. E. (1946). The generalized graphical method for evaluating formation constants and summarizing well field history. *American Geophysics Union Transactions*, **27**, 526–534.

Dardenne M. A. (2000). The Brasília Fold Belt. In: U. G. Cordani, E. J. Milani, A. Thomaz-Filho and D. A. Campos (eds), *Tectonic Evolution of South America*, International Geological Congress, 31. Rio De Janeiro: 231–263.

De Almeida F. F. M., Hasui Y., BritoNeves B. B. and Fuck R. A. (1981). Brazilian Structural Provinces: An Introduction. *Earth-Science Reviews*, **17**, 1–29.

DIN 38412 (L 24); (1981). 38412 Testverfahren mit Wasserorganismen (Gruppe L).

EMBRAPA (1978). Levantamento de reconhecimento dos solos do Distrito Federal. Rio de Janeiro, Boletim Técnico, p. 53.

Faria A. (1995). Estratigrafia e sistemas deposicionais do Grupo Paranoá nas áreas de Cristalina, Distrito Federal e São João D'Aliança - Alto Paraíso de Goiás. Dissertation, University of Brasília, p. 199.

Fiori J. P. O., Campos J. E. G. and Almeida L. (2010). Variabilidade da condutividade hidráulica das principais classes de solos do Estado de Goiás (Variability of hydraulic conductivity of the main soil classes in the Goiás state). *Geosci J*, **29**(2), 229–235.

Freitas-Silva F. H. and Campos J. E. G. (1999). Geologia do Distrito Federal. In: J. E. G. Campos and F. H. Freitas-Silva (eds), Inventário hidrogeológico e dos recursos hídricos superficais do Distrito Federal. SEMATEC: IMEA: MMA-SRH, Brasília. CD-ROM.

Gonçalves T. D. (2012). Recursos Hídricos no Distrito Federal: Modelagem Hidrológica para subsidiar a gestão sustentável na bacia do Ribeirão Pipiripau. Dissertation, Universidade de Brasília. Instituto de Geociências. Brasília, Brazil, p. 134.

Gonçalves T. D., Fischer T., Gräbe A., Kolditz O. and Weiss H. (2013). Groundwater flow model of the Pipiripau watershed, DF of Brazil. *Environ Earth Sci*, **69**, 617–631.

Gonçalves T. D., Lohe C. and Campos J. E. G. (Submitted). Porous aquifers from the DF, Brazil: Characterization and evaluation of hydraulic conditions. *Journal of Soil Science and Environmental Management* - Manuscript Number: JSSEM/09.11.13/0431.

Heitfeld K. H. (1979). Durchlässigkeitsuntersuchungen mittels WD-Test. *Mitt. Ing.- u. Hydrogeol.*, **9**, 175–218.

INMET-National Institute of Meteorology (2009). Automatic stations. Available at http://www.inmet.gov.br/portal/index.php?r=home/page&page=rede_estacoes_auto

Hiltmann W. and Stribrny B. (1998). Handbuch zur Erkundung des Untergrundes von Deponien und Altlasten. Tonminerale und Bodenphysik, Springer Deutschland, p. 1102.

Hirata R. and Conicelli B. P. (2012). Groundwater resources in Brazil: A review of possible impacts caused by climate change. *Anais da Academia Brasileira de Ciências*, **84**(2), 297–312.

IBGE (2002). Instituto Brasileiro de Geografia Estatística. Pesquisa Nacional de Saneamento Básico – 2000. Rio de Janeiro, IBGE, CD-ROM.

Joko C. T. (2002). Hidrogeologia da região de São Sebastião – DF: implicações para a gestão do sistema de abastecimento de água. Dissertação de Mestrado, Instituto de Geociências, Universidade de Brasília, p. 158.

Lousada E. O. (2005). Estudos hidrogeológicos e isotópicos no Distrito Federal: Modelos conceituais de fluxo. Brasilia-DF. Tese de Douturado, Instituto de Geociências, Universidade de Brasília, p. 124.

Lousada E. O. and Campos J. E. G. (2005). Proposta de modelos hidrogeológicos conceituais aplicados aos aquiferos da região do Distrito Federal. *Revista Brasileira de Geociências*, **35**(3), 407–414.

Lousada E. O., and Campos J. E. G. (2011). Estudos isotóticos em água subterrâneas do Distrito Federal: subsídios ao modelo conceitual de fluxo. *Revista Brasileira de Geociências*, **41**(2), 355–365.

Mendoça A. F., Pires A. C. B. and Barros J. G. C. (1994). Pseudo sinkholes occurrences in Brasilia, Brazil. *Environmental Geology*, **23**, 36–40.

Miotliński K., Barry K., Dillon P. and Breton M. (2010). Alice Springs SAT Project Hydrological and Water Quality Monitoring Report 2008–2009. CSIRO Water for a Health Country Research Flagship series, National Water Commission, Australian Government, p. 85.

Moraes L., Santos R. and Souza M. (2008). Monitoramento das águas subterrâneas como instrumento de gestão: o caso da CAESB, DF. In: Congresso Brasileiro de Águas Subterrâneas, 15, Natal. ABAS. CD-ROM.

Mota De Souza M. (2013). Determinação Das Áreas De Recarga Para A Gestão do Sistema Aquífero Físsuro-Cárstico da Região de São Sebastião/DF, Master's Thesis, University of Brasília, p. 83.

Neufeldt H. (2006). Geological drivers of cerrado heterogeneity and 13C natural abundance in oxisols after land-use change. *Rev. Bras. Ciênc. Solo*, **30**(5), 891–900.

Oliveira-Filho A. T. (1992). The vegetation of Brazilian 'murundus' – the island-effect on the plant community. *Journal of Tropical Ecology*, **8**, 465–486.

Pacheco W. L. (2012). Águas subterrâneas do Distrito Federal – Efeito de sazonalidade e características associadas aos isótopos de Deutério, Oxigênio e Carbono. Dissertação de Mestrado, Instituto de Geociências, Universidade de Brasília, p. 129.

PGIRH (2006). Plano de gerenciamento integrado de recursos hídricos do Distrito Federal. PGIRH/DF 1-6. Edited by GDF Brasília. Secretaria de Infra-Estrutura e Obras.

PGIRH/DF (2012). Plano de gerenciamento integrado de recursos hídricos do Distrito Federal – Relatório Síntese. Governo do Distrito Federal, Secretaria de meio Ambiente e Recursos Hídricos do Distrito Federal – SEMARH, Agência Reguladora de Águas, Energia e Saneamento Básico do Distrito Federal – ADAS, p. 98.

Schmelzbach C., Tronicke J. and Dietrich P. (2011). Three-dimensional hydrostratigraphic models from ground-penetrating radar and direct-push data. *J. Hydrol.*, **398**(3–4), 235–245.

Souza M. T. (2001). Fundamentos para Gestão dos Recursos Hídricos Subterrâneos do Distrito Federal - Basics for Management of Groundwater Resources in the DF. Dissertation, University of Brasília, Brasília, p. 94.

Strauch M., Lima J. E. F. W., Volk M., Lorz C. and Makeschin F. (2013). The impact of Best Management Practices on simulated streamflow and sediment load in a Central Brazilian catchment. *Journal of Environmental Management*, **127**, 24–36.

Theis C. V. (1935). The relation between the lowering of the piezometric surface and the rate and duration of discharge of the well using groundwater storage. *American Geophysics Union Transactions,* **16**, 519–524.

Tundisi J. G. (2005). Águas no século XXI: enfrentando a escassez. 2nd Edition, São Carlos: RiMa: IIE, p. 248.

Vienken T., Leven-Pfister C. and Dietrich P. (2012). Use of CPT and other direct push methods for (hydro-) stratigraphic characterization – a field study. *Can. Geotech. J.,* **49**(2), 197–206.

Wild A. (1950). The retention of phosphate by soil. A review. *Journal of Soil Science,* **1**(2), 221–238.

Chapter 4

Land use management as part of Integrated Water Resource Management

C. Lorz, C. Franz, L. Koschke, F. Makeschin and M. Strauch

4.1 INTRODUCTION

The demand for integrating land use in IWRM concepts has recently increased worldwide. This might be also due to the fact of rising awareness for sustainable use of water resources and the pressure from society for a reliable and healthy water supply. This applies especially for regions where (temporary) water scarcity, water quality problems and intensification of land use are issues, as it is the case in Central West Brazil. The project IWAS-ÁGUA DF identified four major aspects for this region focusing on land use management and IWRM, (1) analysis of land use history; (2) simulation of effects of land management practices; (3) development of a planning support tool; and (4) identifying sediment sources.

The topics 1–3 have been dealt with focus on the 215 km^2 meso-scale catchment Pipiripau, in the northeastern DF. Mean discharge of the Pipiripau river is around 3 m^3 s^{-1} for the period 1971–2009. Land use is predominantly agriculture with a share of 60%. Water is extracted from rivers and creeks for irrigation and, additionally, at the pumping station Montante Captação (Figure 4.1) for drinking water supply of the nearby town Planaltina. The Pipiripau River Basin is part of the Brazilian program *produtor de água* (BRASIL, 2010; www.ana.gov.br), which aims at improving water quality and quantity by restoring or preserving (semi) natural vegetation along streams and by implementing Best Management Practices in agriculture. This project aims to introduce Payment-for-Environmental-Services schemes and to support the dialogue between stakeholders.

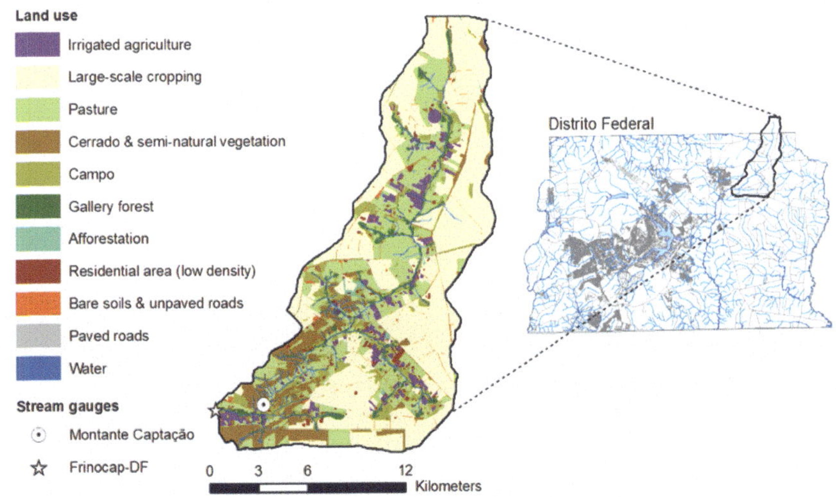

Figure 4.1 Location and land use of the Pipiripau River Basin.

Topic 4 was investigated in the catchment of the Lago Paranoá reservoir (Figure 1.2), which is an artificial lake with a surface area of about 38 km , formed by the construction of the Paranoá Dam in 1962. The size of Lago Paranóa catchment is 1046 km . Its main tributaries are several small rivers, including the Riacho Fundo Creek and Gama Creek in the south, and the Bananal and Torto Creek in the north. The catchment of the Lago Paranoá has experienced highly dynamic land use since 1960, and has undergone a substantial shift from natural vegetation cover to urban areas and to agricultural use (Lorz *et al.* 2011; Menezes, 2010). Urban land use, including residential, commercial, and infrastructural development, occupies approximately 34% of the catchment area. Only 8% is used for agricultural production and 58% of the land is in (semi) natural state.

4.2 ANALYSIS OF EFFECTS OF LAND USE HISTORY ON WATER RESOURCES IN THE PIPIRIPAU RIVER BASIN

The effects of land use/land cover change (LULCC) on surface water are best shown for the Pipiripau river basin which experienced a strong expansion of cropland since 1960. For the main tributary (Pipiripau river) a statistically significant decrease of base flow discharge (5th percentile of discharge, Figure 4.2) since 1979 has been observed. This trend seems to prove true for other river basins in DF, which show a decrease of 40–70% during the last three decades. Since only small trends in amount and patterns of rainfall have been observed, we assume LULCC to be the main driver for decreasing base flow discharge in both river basins (Figure 4.3; Lorz *et al.* 2011). A similar relationship, but with an increase of mean discharge and higher rainy season discharge, was found by Costa *et al.* (2003) for the Tocantins River Basin in Central Brazil.

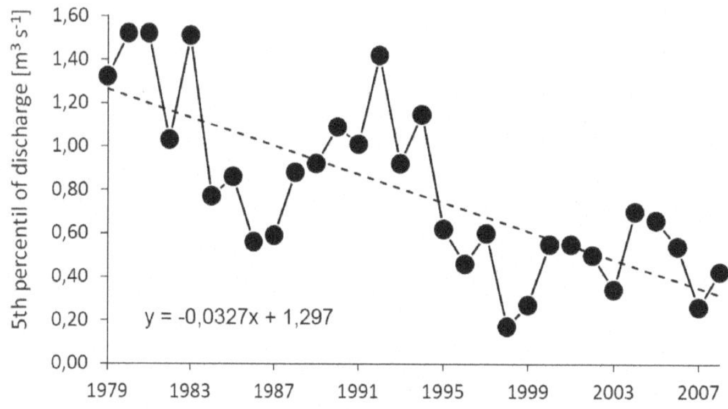

Figure 4.2 Time series of 5th percentile of annual discharge and trend line for the Pipiripau River Basin; Stream gauge Frinocap – DF (see Figure 4.1 for location), code 60473000, A = 215 km^2, cropland 80.6 % (Lorz *et al.* 2012).

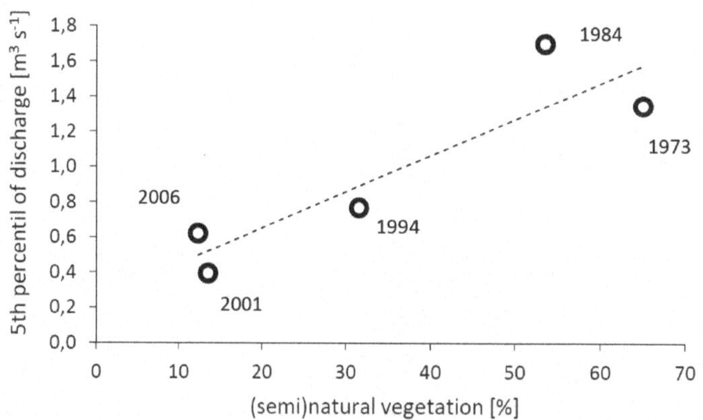

Figure 4.3 (Semi)natural vegetation cover and discharge (base flow, 5th percentile) for the Pipiripau River Basin, percentage of land use classes are for the years 1973, 1984, 1994, 2001 and 2006, 5th percentile (run off) are means for the years 1973–1975, 1984–1986, 1994–1996, 2001–2003 and 2006–2008 (Lorz *et al.* 2012).

Two major processes might explain this phenomenon. (1) Higher evapotranspiration of cultivated crops will use more soil water than (semi)natural vegetation. Thus, groundwater recharge and groundwater flow into surface water is reduced, that is, base flow discharge is lower. (2) Water extraction from surface and ground water for irrigation purposes has increased along with the expansion of cropland (Lima *et al.* 2004; Oliveira & Talamini, 2010). Irrigated areas accounted for 11,227 ha in 1998 and 16,039 ha in 2004, that is, increase by 43%, for the wider region of Brasília (PGIRH, 2006).

Since there is a strong linkage between land use and water quality we expect the substantial LULCC in the DF to have also severe effects on water quality parameters. We analysed the effects of size of river basin and shares of urban, agricultural and forested areas on chemical oxygen demand (COD), ammonium concentration (NH_4^+) and turbidity for 17 river basins in DF (Lorz *et al.* 2011). Between size of river basin and COD, NH_4^+ (Figure 4.4) and turbidity positive correlations were found. Based on the spatial patterns of settlements we assumed for COD and NH_4^+ that the number of sources for pollutants and the concentration in stream flow increases with the size of river basins. Smaller river basins are most likely to be headwaters with less pollution sources. The increase of turbidity with river basin size might also be explained by remobilization of alluvial sediments by lateral erosion depending on the length of the river and the thickness of sediments, both increasing with river basin size. The share of urban areas has a very strong effect on water quality of surface waters. COD and NH_4^+ concentrations increase with the area of settlements in the river basin as it was also found by other studies in Brazil (Conceição *et al.* 2010; Hepp & Santos, 2009; Menezes *et al.* 2009; Zeilhofer *et al.* 2006). However, surface water from rural areas in Brazil might be sometimes more polluted than urban rivers (Kühl *et al.* 2010). Future aims of our research are to identify sources of pollution, that is, effects of horticulture and meat production (Bilich 2007) or the role of afforestation and riparian zone (Chaves & Santos, 2009).

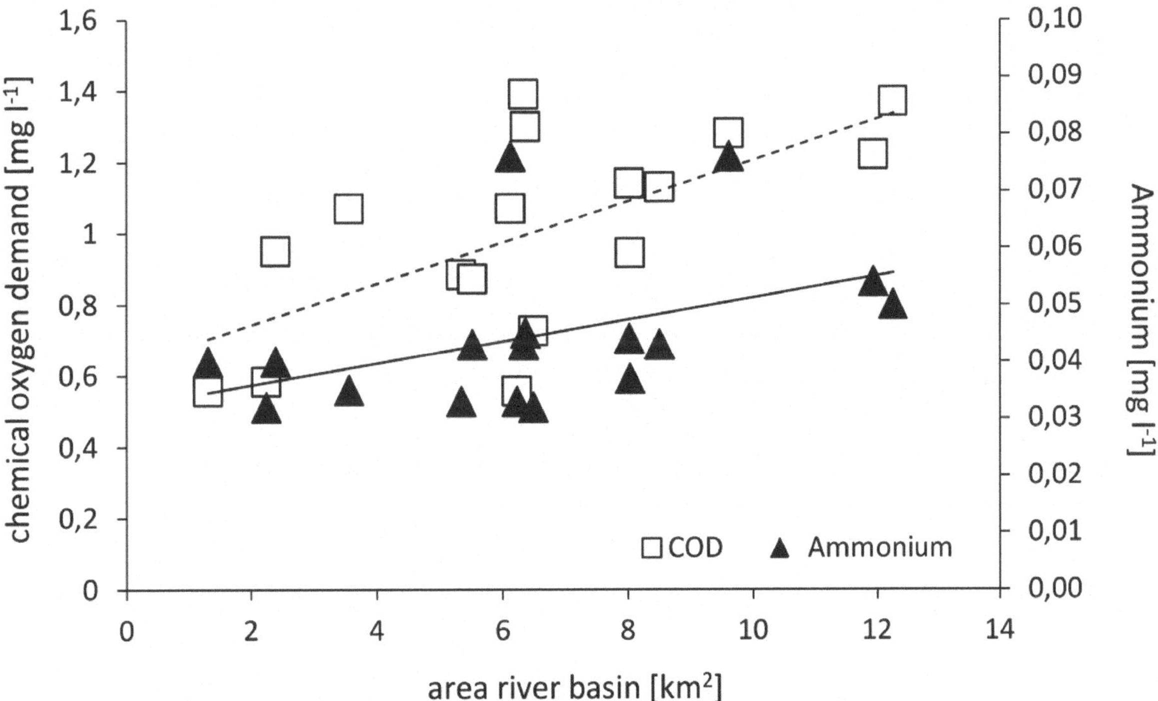

Figure 4.4 Area of river basin and water quality, that is, chemical oxygen demand (COD) and ammonium (NH_4^+) (Lorz *et al.* 2012).

4.3 USE OF SIMULATION MODELS FOR ASSESSING LAND MANAGEMENT PRACTICES – EXAMPLE PIPIRIPAU RIVER BASIN

Watershed models have become a main tool for problems related to water resources assessment and management (Singh & Frevert, 2006; Daniel *et al.* 2011; Jajarmizadeh *et al.* 2012). They are used to simulate bio-physical processes of the flow of water, sediment, chemicals, nutrients, and microbial organisms within watersheds, as well as to quantify the anthropogenic impact on these processes (Singh & Frevert, 2006). In order to study the effects of land management practices on catchment hydrology and sediment transport in the Pipiripau River Basin, whose water resources are intensively used for both agriculture and human water supply, the Soil and Water Assessment Tool (SWAT; Arnold *et al.* 1998; Arnold & Fohrer, 2005) was utilized following the workflow presented in Figure 4.5.

One of the most challenging issues faced by watershed modelers in tropical regions is the fact that regular rain gauge networks usually do not reflect the high spatio-temporal variability of predominantly convective precipitation patterns. Therefore, an ensemble of different reasonable input precipitation data-sets differing in terms of data source (ground or satellite based), spatial distribution (lumped or distributed), and temporal distribution (raw time series or moving average) was used to examine the uncertainty in parameterization and model output (Strauch *et al.* 2012). Plausible streamflow and sediment load predictions could be achieved for each input data-set (Figure 4.6.). However, the best-fit parameter values after model calibration varied widely across the ensemble. Due to its enhanced consideration of parameter uncertainty, this ensemble approach provides more robust predictions and hence is reasonable to be used also for scenario simulations.

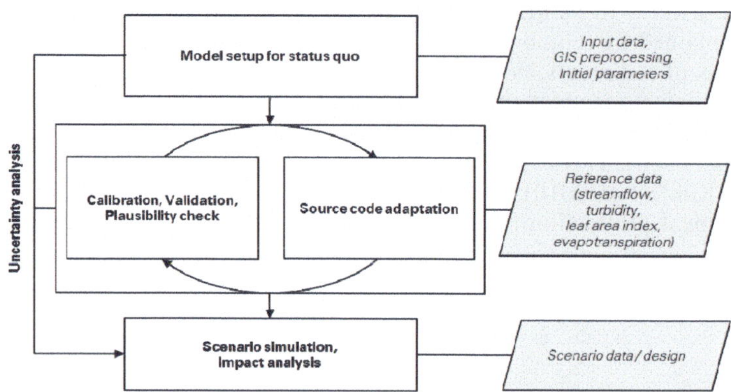

Figure 4.5 Modeling workflow.

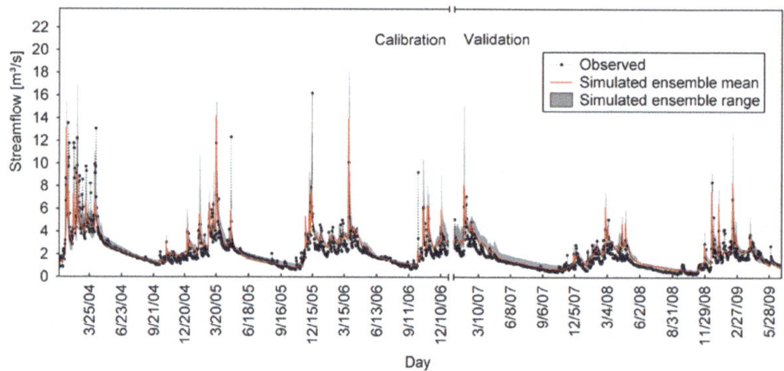

Figure 4.6 Simulated vs. observed streamflow at gauge Montante Captação, the ensemble range refers to the range of simulations based on different precipitation inputs (Strauch *et al.* 2013).

Scenarios considering Best Management Practices (BMP) for the Pipiripau River Basin showed that erosion control structures, such as terraces and small retention basins along roads (barraginhas) are promising measures to reduce sediment loads (up to 40% in combination) while maintaining streamflow (Strauch *et al.* 2013; Figure 4.7). Tests for a multi-diverse crop rotation system, in contrast, showed a high vulnerability of the hydrologic system against any increase in irrigation in form of substantially declining low flows during dry season (Figure 4.7a). Considering the BMP implementation costs, it was possible to estimate cost-abatement curves (Figure 4.7b), which can provide useful information for watershed managers, especially when BMPs are supported by Payments for Environmental Services as it is the case in the study area due to the program *produtor de água* (BRASIL, 2010). As an example, a one-million-USD investment for implementing Barraginhas was predicted to reduce average sediment loads of the Pipiripau River by 18 to 26% depending on which rain input model was used. This range is similar to the predicted effect of terraces (16–23%). However, while this was the maximum effect for Barraginhas (assuming the extreme scenario with ten Barraginhas per road kilometer), the maximum effect for terraces (when installed on total farmland) was a sediment load reduction of up to 37% for an investment of approx. 1.75 million USD. The strongest reduction of average sediment load was predicted for a combined implementation of Barraginhas and terraces (each in maximum), ranging from 34 to 48%, but this was related with substantially decreased cost-effectiveness (2.75 million USD).

While for agricultural areas the model has proven to generate plausible results, the plant growth module of SWAT was found to be not suitable for simulating perennial tropical vegetation, such as Cerrado (Brazilian Tropical Savanna) or Mata Galeria (gallery forest), which can also play a crucial role in river basin management. For temperate regions SWAT uses dormancy to terminate growing seasons of trees and perennials (Neitsch *et al.* 2011). However, there is no mechanism considered to reflect tropical seasonality, that is, the phenological change between wet and dry season. Therefore, a soil moisture based approach was implemented in the plant growth module to trigger new growing cycles in the transition period from dry to wet season (Strauch & Volk, 2013). The adapted model was successfully tested against LAI (Figure 4.8) and ET time series derived from remote sensing products (MODIS; Myneni *et al.* 2002; Mu *et al.* 2011). Since the proposed changes are process-based but also allow flexible model settings, the modified plant growth module can be seen as a fundamental improvement useful for future model application in the tropics.

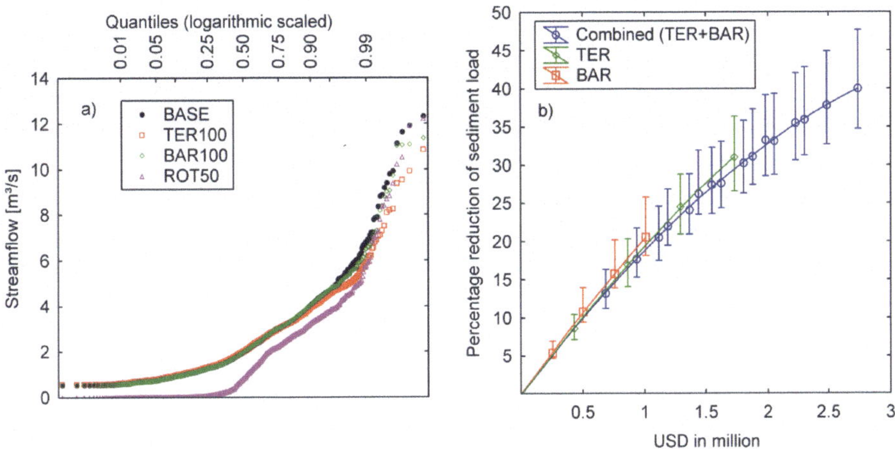

Figure 4.7 (a) Quantile plot of simulated daily streamflow at gauge Montante Captação (ensemble mean, period 2004–2009) for the baseline scenario without BMP implementation (BASE) and conservation management scenarios (TER = terraces, BAR = Barraginhas, ROT = multi-diverse crop rotation), each with maximum intensity; (b) Average percentage reduction of sediment load due to different BMP strategies related to the costs of implementation (curves were derived by polynomial fitting), note that the range shown as whiskers is due the use of different rain input models (Strauch *et al.* 2013, modified).

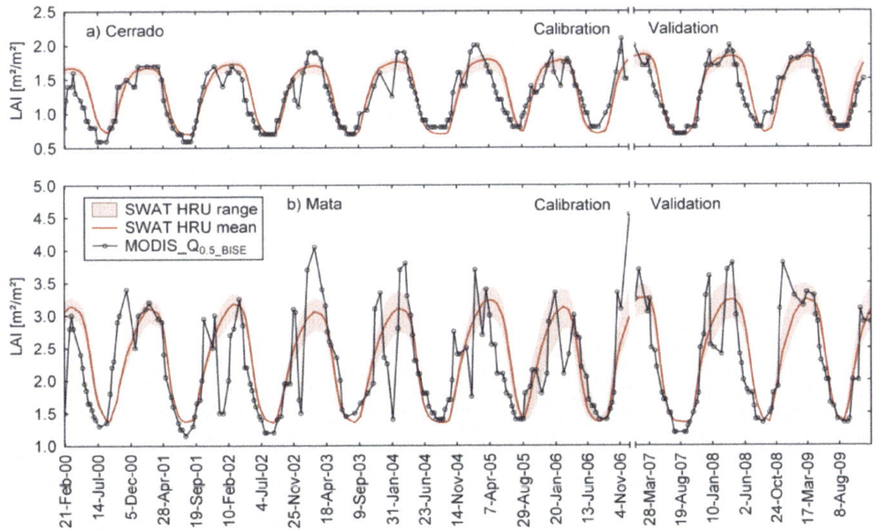

Figure 4.8 Simulated Leaf Area Index (LAI, mean and range over spatial model units – HRU) for vegetation type (a) Cerrado and (b) Mata compared to the corresponding MODIS reference data (Strauch & Volk, 2013).

The SWAT case studies (detailed information in Strauch *et al.* 2012, 2013; Strauch & Volk, 2013) show insights into the workflow of a watershed model application in the semi-humid tropics – from input data processing and model setup over source code adaptation, model calibration and uncertainty analysis to its use for running scenarios. They depict region-specific challenges but also provide practical solutions and thus further steps toward robust and process-based model predictions to assist land and water resources management.

4.4 *LETSMAP DO BRASIL* – A WEB-BASED PLANNING SUPPORT TOOL FOR SEDIMENT MANAGEMENT IN THE PIPIRIPAU RIVER BASIN

Urban sprawl and intensive agriculture in Central West Brazil led to large scale degradation and loss of Cerrado, the native savannah vegetation (Schmidt *et al.* 2009). Current land use causes high loads of suspended solids in stream water and subsequent silting of drinking water reservoirs (Felizola *et al.* 2001; ANA, 2004). These processes threaten the regional water resources by deterioration of (raw) water quality and reduction of the volume of reservoirs (Fortes *et al.* 2007; Lorz *et al.* 2011).

Landscape processes and functions (LPF), important for humans, are frequently related directly or indirectly to water resources. LPF that are closely related to the hydrological cycle and land cover/land use in a watershed (Pattanayak & Wendland, 2007) are for instance sediment retention, runoff control, and nitrogen retention (Lorz *et al.* 2013). The restoration and conservation of natural vegetation in priority areas such as riparian zones has been proposed to reduce to soil erosion and nutrient runoff from adjacent agricultural areas (*produtor de água*, www.ana.gov.br).

In order to support Integrated River Basin Management (IWRM) in the Pipiripau river basin, we use the planning support tool Letsmap do Brasil (Lorz *et al.* 2013) that was developed within the GISCAME framework (Fürst *et al.* 2010; Fürst *et al.* 2013; Koschke *et al.* 2012). Letsmap do Brasil is a raster based platform which enables the impact assessment of LULCC scenarios on land use planning targets, which can be expressed for instance by LPF, landscape services, ecosystem services or others (Fürst *et al.* 2010). It allows to run land use/land cover change scenarios to assess the impact on LPF following simulated LULCC und thus to support sediment management and trade-off analyses between different LPF.

The main information layer is the raster based land use/land cover map. The core of the system is the option for the user to change directly the land use/land cover of single cells or patches of cells and to derive immediate feedback of the effects on LPF. To each of the land use/land cover classes a relative, semi-quantitative value (0–100 points) is assigned in a first assessment step. A value of 100 would mean the maximum regional LPF whereas a value of 0 indicates zero performance of the respective LPF (Fürst *et al.* 2010; Koschke *et al.* 2012). Using the basic land use/land cover type values, Letsmap do Brasil calculates the area weighted mean according to the share of all land use/land cover types that can be found in the study area. The assessment results of LULCC scenarios are provided as feed-back to the user in form of a star diagram or tables. Additionally, the current values assigned to each land use/land cover can be visualized in order to support a mapping of LPF.

To direct restoration and conservation efforts on high value areas and to allow an ex-ante impact assessment of land use change scenarios, physical characteristics of watersheds such as topography, soil quality and location need to be considered (Rounsevell *et al.* 2012). We implemented GIS based analyses to identify areas with differing landscape properties and potentials (LPP) which will impact the land use based assessment in order to allow the inclusion of site specific conditions in addition to the land use specific assessment approach of Letsmap do Brasil. A further aim was to identify priority areas that need special attention from land managers and are very effective in delivering the desired LPF if applied appropriately. We ensured transferability to other regions and intended to constrain time demand for the set-up of the system by using basic, easily available environmental data.

The Letsmap assessment approach includes three main steps, (1) assessment of the land use type specific contribution to LPF; (2) Estimation of the LPP that are related to LPF; and (3) Spatially explicit assessment of LPF through combination of results from step 1 and step 2 (Figure 4.9).

The assessment of land use types contribution to LPF was based on the indicators nitrogen load (applied fertilizer in kg ha^{-1} a^{-1}) for nitrogen loss control, C-Factor (USLE) for sediment retention, CN value for runoff control, and yield (Mg ha^{-1} a^{-1}) and market price (€ ha^{-1}) for the agronomic value. All values were standardized to a scale ranging from 0-100 points (step 1). Besides the land use map, the LPP layer represents a secondary information layer and was introduced in Letsmap do Brasil. It relates site specific conditions to LPF. For each LPF, a respective layer showing the related LPP was provided (Figure 4.9, step 2). The secondary layer enables a spatial explicit assessment of LPFs by taking the properties and potentials of a specific site (cell) into account to provide the selected LPF. This is done through combination of the land use related assessment (step 1) with LPP (step 2). For each LPF assessment approaches were applied to identify related LPP. The methodical approach has been operationalized for nitrogen loss control, sediment retention, runoff control, and agronomic value (see also Lorz *et al.* 2013). In order to explain the methodological approach in detail we focus on the assessment methodology for nitrogen loss control.

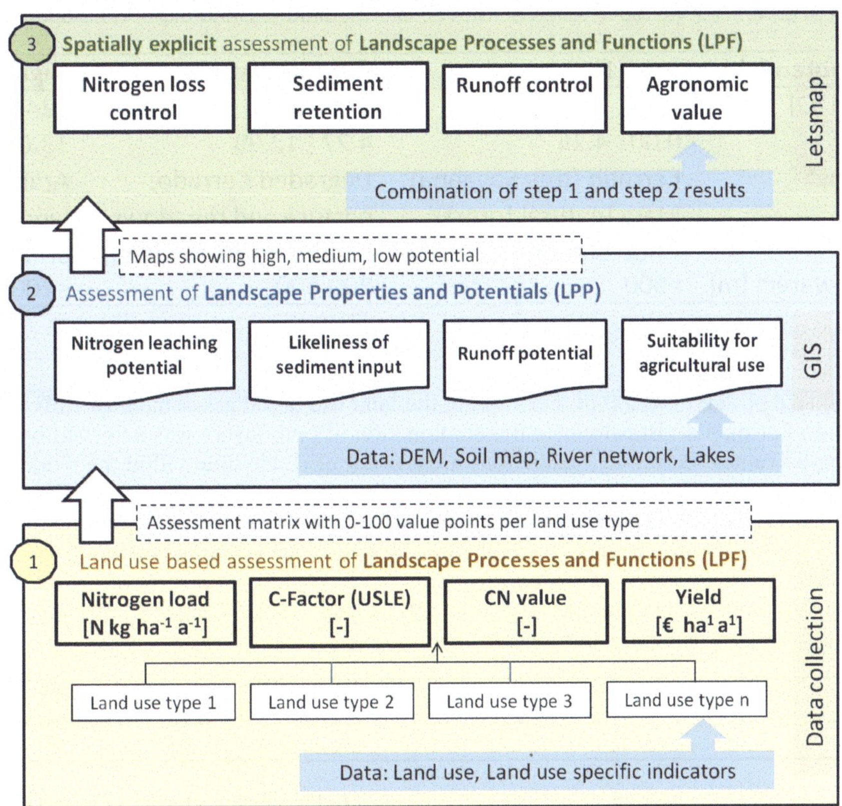

Figure 4.9 Overview of the assessment steps, by combining land use related LPF with LPP the assessment becomes spatially explicit.

For step 1, information on the average amount of nitrogen fertilizer applied within each land use type was collected. The values for this indicator were taken from Lorz *et al.* (2013), IBGE (2006) and EMBRAPA (2012). Beside land use which impacts leaching of nitrogen into adjacent water bodies, biophysical properties influence the potential of a landscape to retain nitrogen. For step 2, an approach of qualitative estimation of the nitrogen leaching potential was adapted from Orlikowski *et al.* (2011). Required input parameter are land use, soil (root zone available water capacity, RZAWC), slope, riparian buffer strips, and distance to surface waters (Figure 4.10). For soil parameters we used data of 16 soil profiles which were the basis for a regionalization to the study area (Strauch *et al.* 2013).

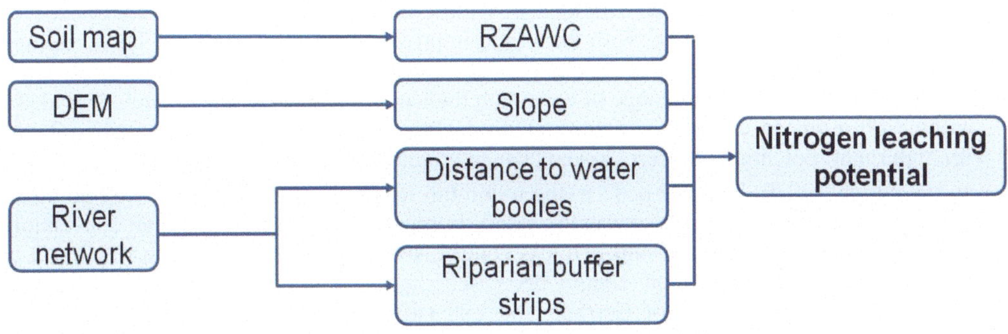

Figure 4.10 Flowchart showing the steps to calculate the nitrogen leaching potential (adapted from Orlikowski *et al.* 2011).

We used ArcMap10 to compile the data and to finally classify the parameters as well as the LPP map that resulted from the overlay into the three classes (1) low, (2) intermediate, and (3) high nitrogen leaching potential. Classes 1–3 were assigned using individual value ranges observed in the study area and taking into account the influence of parameter values on this LPP (Table 4.1).

Table 4.1 Classification of parameter values in terms of low (1), intermediate (2), and high (1) nitrogen leaching potential.

Nitrogen leaching potential	Low (1)	Intermediate (2)	High (3)
RZAWC total [mm H2O]	<142	142-244	>244
Slope [Degree]	0.00 - 4.26	4.27 - 12.99	13.00 - 60.43
Riparian buffer strips*	Cerrado (tree Savanna), Mata (natural forest), Afforestation	Degraded Cerrado, pasture and meadows, Campo (grass savanna)	Arable land, Irrigated land, Bare soil, Build up areas
Distance to surface waters [m]	>800	200-800	<200

*within 200 m distance (minimum buffer width) to surface waters

We used a raster cell size of 200 m × 200 m. For linking the land use based assessment of LPF with LPP, we uploaded the maps as grid files into Letsmap do Brasil where the combination of both layers was achieved by percentage reductions of the initial land use based values. Percentage reductions are a result of discussion within the working group. Cells with a high nitrogen leaching potential are therefore subject to a higher reduction of value points in comparison to cells with a low nitrogen leaching potential (Table 4.2). The highest reduction occurs if land use types with a high fertilizer input and thus with a low value for nitrogen loss control occurs in a cell with a high nitrogen leaching potential.

Table 4.2 Connection matrix indicating the percentage reduction of land use type related value points for the LPF nitrogen loss control with respect to the LPP nitrogen leaching potential.

Land use types	LPF Value classes	LPP Low 1	Moderate 2	High 3	
Coffee	0-10	0	-10	-20	
-	11-20	0	-10	-20	
Arable land, cotton;	21-30	0	-10	-20	
Irrigated land	31-40	0	-5	-10	
-	41-50	0	-5	-10	
Vegetables; Fruits	51-60	0	-5	-10	Reduction in % of initial value
Arable land, corn tillage/no tillage	61-70	0	0	-5	
Arable land, general, no tillage	71-80	0	0	-5	
Arable land, bean	81-90	0	0	-5	
Cerrado; Campo; Arable land, sorghum; tillage/ no tillage	91-100	0	0	0	

LULCC scenario simulations using Letsmap do Brasil can be conducted in an explorative manner or by taking reference maps indicating for instance protected areas, areas with a priority for a certain development or risk areas. We tested LULCC scenarios on the basis of the developed LPP maps with the aim to improve nitrogen loss control. Three exemplary scenarios have been run to demonstrate the functioning of Letsmap do Brasil, (1) change of areas with high nitrogen leaching potential into natural savannah (Cerrado); (2) change of cells with the land uses types irrigated land, degraded Cerrado, bare soil and of cells within 200 m distance from surface waters into Cerrado; (3) changes as in scenario 2 and additionally change of areas with high runoff potential into Cerrado and change of arable land into pasture.

In comparison to the current (2010) land use pattern, the minimum restoration scenario 1 showed no difference (Figure 4.11) as the number of cells with changed land use was too small. In scenario 2, it can be seen that conversion into Cerrado in riparian zones led to a significant increase of nitrogen loss control, sediment retention and runoff control. However, due to the reduced portion of arable land the agronomic value decreased slightly. The maximum scenario 3 exemplarily shows the positive effect of large scale restoration of Cerrado and extensification of arable land. While sediment retention increased from 83 to 88 points, and runoff control from 46 to 74 points, nitrogen loss control even reached the maximum value of 100 points compared to 88 in the initial land use pattern. The increase in LPF with focus on sediment generation and water quality is accompanied by a strong reduction of the agronomic value from 30 to 9 points.

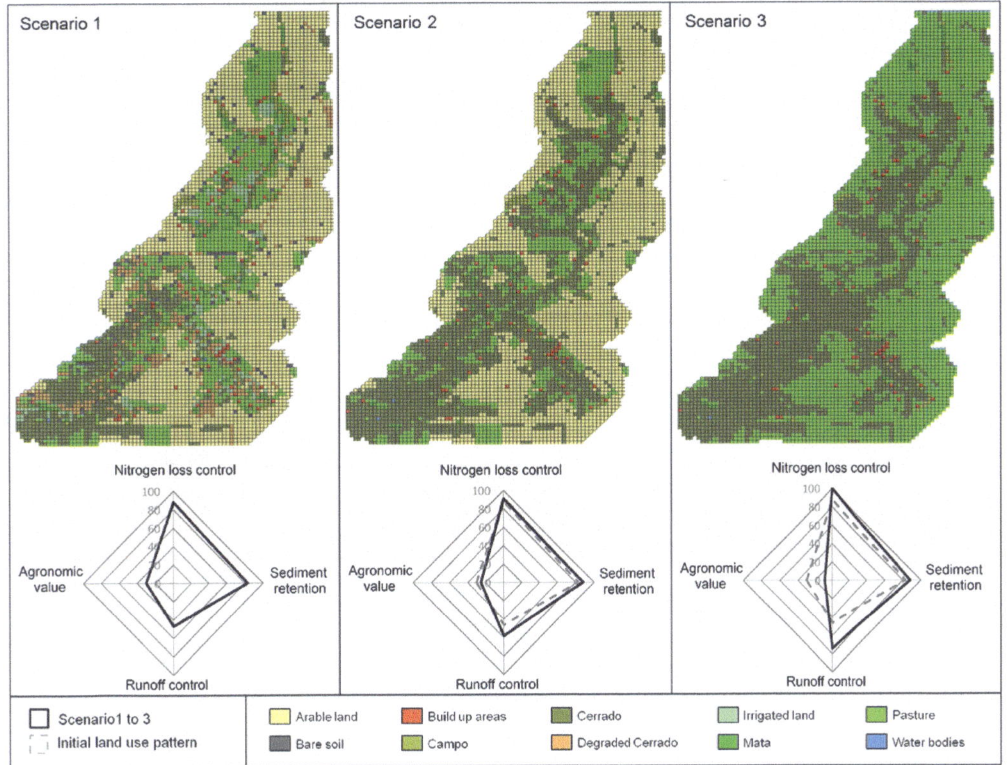

Figure 4.11 Screenshots of Letsmap do Brasil showing the results of scenarios (black line in star diagram) changes and their impacts on LPF compared to the initial land use pattern (dotted grey line). Assessment values in star diagrams range between 0-100 points. Scenarios: 1 – Areas wit h high nitrogen leaching potential converted into Cerrado, 2 – Conversion of irrigated land, degraded Cerrado, bare soil and of cells within 200 m distance from surface waters into Cerrado, 3 – Conversion as in 2 and change of areas with high runoff potential into Cerrado and of arable land into Pasture (Koschke *et al.* 2014).

Areas where a low nitrogen leaching potential was identified are currently covered by (semi) natural vegetation such as Cerrado, Campo, and Mata. Thus, the effects of LULCC as in scenario 1 have a marginal impact on assessment results. Yet, it is now possible to take site specific properties and potentials into account which facilitates trade-off analyses between LPFs. Our findings suggest that further degradation and loss of native land use types must be avoided in order to safeguard water resources in the Pipiripau river basin.

Assessment results, especially in terms of sediment retention, may reflect the problems of degraded water resources (CAESB, 2001; Felizola *et al.* 2001; Fortes *et al.* 2007) only with restrictions. In the study region, water resources might be threatened also by land use types which will be not detected at meso-scale level because of their spatial extent or temporary character. Franz *et al.* (2014) showed that sediment input in the Lago Paranoá comes mostly from urban point, e.g., construction sites or unpaved roads. Therefore, small scale erosion events and point sources of sediments and pollution can be addressed only with restrictions given the available land use data and assessment approach. In addition, LULCC might have unforeseen (nonlinear) and ambiguous effects and land use related indicator values might show widely varying ranges. Since, the linking of LPF with LPP is based on a subjective evaluation, uncertainty is very difficult to quantify (for an extended discussion see Koschke *et al.* 2013).

A major advantage of Letsmap do Brasil is the inclusion of environmental data which is the base for the spatially explicit evaluation of LPF. In situations where data access/availability and/or resources to process or collect data are rather limited, the presented approach may help to conduct an estimation of regional potentials and potential threats. The accuracy and practical relevance of applying the tool in IWRM and planning could be further enhanced by a methodical refinement which relates to the use of more detailed information on land use and environmental parameters with respect to both, thematic and spatial resolution. Letsmap do Brasil approach might be applied for example in the frame of the project *produtor de água* (www.ana.gov.br) which aims to improve water quality and quantity through restoration and preservation of (semi) natural vegetation along rivers. Major general advantages of the tool are its easy accessibility, low computational requirements as a result of the client server architecture and the resulting fast calculation of results. Due to easy handling and a less complex modeling approach it is a user-friendly system to visualize LULCC and to inform non-experts on the effects of LULCC (participation). The underlying layer and LPF distribution maps which can be displayed in Letsmap do Brasil can be used as a starting point for prioritization of areas. The tool hence can promote the dialogue between stakeholders and supports integrated land use planning.

4.5 SEDIMENT SOURCES AND SEDIMENT MANAGEMENT

Siltation of reservoirs is a major challenge in the study region. The consequences of siltation might be the total loss of the storage capacity for smaller reservoirs and a loss of storage capacity for bigger reservoirs. In addition, sediment bodies accumulate potential pollutants and nutrients, which might be released under changing hydrochemical regimes in reservoirs.

For a better understanding of sediment generation we used the idea of the sediment cascade (Figure 4.12). In the catchment of the Lago Paranoá four major sources were found, where sediments are mobilized and transported towards the final sinks, that is, lake/reservoir. The slopes of agricultural land, urban areas/construction sites and gullies (Figure 4.13) are major sources of sediments outside alluvial plains. The alluvial system itself might be a major source of sediments and temporal sink at the same time. Final sinks are the alluvial systems in the inflow of reservoirs and the reservoirs. In addition, a great number of small basins for sediment retention (*barraginhas*) have been built (Figure 4.13), but in the long-run these sinks are rather temporary with varying life time depending on the level of maintenance.

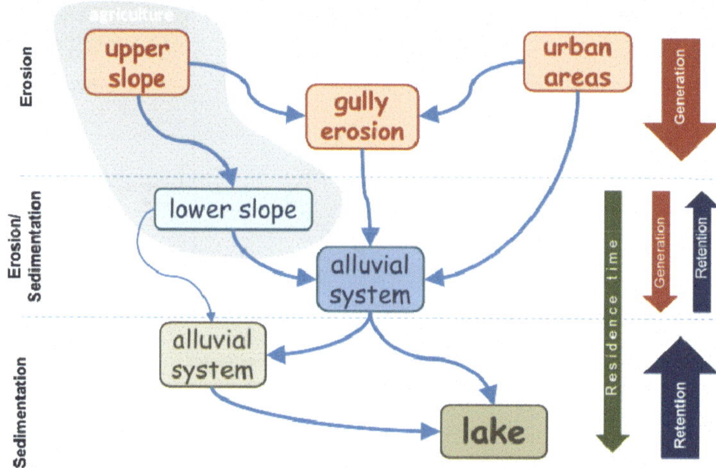

Figure 4.12 Sediment cascade for the Distrito Federal (Franz *et al.* 2014).

Figure 4.13 Sediments in the study region, construction sites near the Riacho Fundo (upper left), gully erosion (upper right), sand banks at the Riacho Fundo (lower left) and sediment micro basins (*barraginhas*) (lower right).

A crucial aspect in understanding the sediment system is the identification and quantification of sediment sources. The sedimentation rate for alluvial flood plains during the period after 1960 have been found to be 10 to 100 times higher compared to periods without human interferences (Franz *et al.* 2012). The identification and quantification of sediment sources have been achieved by using a combined approach of geochemical fingerprinting and statistical analysis (Franz *et al.* 2013, 2014) for the catchment of the Lago Paranoá and the Riacho Fundo sub-catchment (Figure 4.1). By using this approach, urban areas were identified as major sources of alluvial sediments in both catchments, showing a mismatch between spatial share and contribution to alluvial sediments (Table 4.3.). A further distinction of urban areas in the Riacho Fundo sub-catchment showed that sediment generation in developed urban areas with paved roads is limited. In contrast, the sediment delivery rate from still expanding urban areas with extensive construction activities, semi-detached housings, and unpaved roads is rather high. Bare soils and exposed grounds are the 'hot spots' of sediment generation. Three quarter of the alluvial sediments come from construction sites and less developed residential areas in the Riacho Fundo sub-catchment.

Table 4.3 Results of the sediment source contribution for the whole Lago Paranoá catchment and Riacho Fundo sub-catchment (Franz *et al.* 2013).

Source category	Weighted mean relative sediment contribution* (%)	Share of land use (%)
Catchment Lago Paranoá		
Urban	85 ± 4	34
Agricultural	5 ± 4	8
Natural	10 ± 2	58
Riacho Fundo sub-catchment		
Urban	53 ± 4	38
Agricultural	31 ± 2	47
Natural	16 ± 3	12

*Relative mean errors for Lago Paranoá 13.9% and Riacho Fundo 12.3%.

The small share of agricultural areas in the Lago Paranoá basin is reflected by the low contribution to alluvial sediments (Table 4.3). In the Riacho Fundo sub-catchment the larger share of agricultural area is underrepresented in the alluvial sediments (Table 4.3). Natural areas do not substantially contribute to alluvial sediments in the Lago Paranoá basin despite of their high spatial share. The relative higher share of sediments from natural areas in the Riacho Fundo sub-catchment might be explained by higher shares of bare soils which favour the formation of gullies which may expand uphill and thus also affect natural areas.

4.6 REFERENCES

ANA (Agência Nacional de Águas) (2004). Estudos de disponibilidade de agua para a Bacia do Pipiripau, Brasília, DF.

Arnold J. G. and Fohrer N. (2005). SWAT2000: Current capabilities and research opportunities in applied watershed modelling. *Hydrol. Processes*, **19**, 563–572.

Arnold J. G., Srinivasan R., Muttiah R. and Williams J. (1998). Large area hydrologic modeling and assessment part I: Model development. *J Am. Wat Res Ass*, **34**, 73–89.

Bilich M. R. (2007). Ocupação das terras e a qualidade da água na microbacia do Ribeirão Mestre d'Armas, Distrito Federal. M. Sc. thesis, Universidade de Brasília, Brazil.

BRASIL (2010). Programa Produtor de Água - Relatório de Diagnóstico Socioambiental da Bacia do Ribeirão Pipiripau, Brasília, DF.

CAESB (Companhia de Saneamento Ambiental do Distrito Federal) (2001). Plano de Protecao Ambiental do Ribeir o Pipiripau, Brasília, DF.

Chaves H. M. L. and Santos L. B. (2009). Ocupação do solo, fragmentação da paisagem e qualidade da água em uma pequena bacia hidrográfica. *Rev. Bras. Engenharia Agrícola Ambiental*, **13**(6), 922–930.

Conceição F. T., Sardinha D. S., Souza A. D. G. and Navarro G. R. B. (2010). Anthropogenic Influences on Annual Flux of Cations and Anions at Meio Stream Basin, São Paulo State, Brazil. *Wat. Air Soil Poll.*, **205**, 79–91.

Costa M. H., Botta A. and Cardille J. A. (2003). Effects of large-scale changes in land cover on the discharge of the Toncantins River, Southeastern Amazonia. *J. Hydrology*, **283**, 206–217.

Daniel E. B., Camp J. V., Leboeuf E. J., Penrod J. R., Dobbins J. P. and Abkowitz M. D. (2011). Watershed Modeling and its Applications: A State-of-the-Art Review. *Open Hydrology J.*, **5**, 26–50.

EMBRAPA (Empresa Brasileira de Pesquisa Agropecuária) (2012). Brazilian agriculture development and changes, www.cecat.embrapa.br, (accessed 22 January 2014).

Felizola E. R., Lago F. P. d. L. S. and Galvão W. S. (2001). Avaliação da dinâmica da paisagem no Distrito Federal. Projeto da Reserva da Biosfera do Cerrado e Fase I. In: Anais X Simpósio Brasileiro de Sensoriamento Remoto, Foz do Iguaçu, 21–26 April 2001. INPE, Foz do Iguaçu, 1593–1600.

Fortes P. T. F. O., Oliveira G. I. M., Crepani E. and Medeiros J. S. D. (2007). Geoprocessamento aplicado ao planejamento e gestão ambiental na Área de Proteção Ambiental de Cafuringa, Distrito Federal Parte 2: processamento de dados espaciais. In: Anais XIII Simpósio Brasileiro de Sensoriamento Remoto, Florianópolis, Brasil, 21–26 April 2007. INPE, Florianópolis, 2613–2620.

Franz C., Roig H., Makeschin F., Schubert M., Weiss H. and Lorz C. (2012). Sediment characteristics and sedimentation rates of a small river in Western Central Brazil. *J Environ. Earth Sc.*, **65**(5), 1601–1611.

Franz C., Makeschin F., Weiss H. and Lorz C. (2013). Geochemical signature and properties of sediment sources and alluvial sediments within the Lago Paranoá catchment, Brasilia DF: A study on anthropogenic introduced chemical elements in an urban river basin. *Sci. Total Environ.*, **452–453**, 411–420.

Franz C., Makeschin F., Weiss H. and Lorz C. (2014). Sediments in urban river basins: Identification of sediment sources within the Lago Paranoa catchment, Brasilia DF, Brazil – using the fingerprint approach. *Sci. Total Environ.*, **466–467**, 513–23.

Fürst C., Frank S., Witt A., Koschke L. and Makeschin F. (2013). Assessment of the effects of forest land use strategies on the provision of ecosystem services at regional scale. *J. Environ. Man.*, **127**, 96–116.

Fürst C., Volk M., Pietzsch K. and Makeschin F. (2010). Pimp Your Landscape: A Tool for Qualitative Evaluation of the Effects of Regional Planning Measures on Ecosystem Services. *Environ. Man.*, **46**, 953–968.

Hepp L. U. and Santos S. (2009). Benthic communities of streams related to different land uses in a hydrographic basin in southern Brazil. *Environ. Monit. Assess.*, **157**, 305–318.

IBGE (Instituto Brasileiro de Geografia e Estatística) (2006). Censo Agropecuário 2006, http://www.ibge.gov.br/, (accessed 24 July2013).

Jajarmizadeh M., Harun S. and Salarpour M. (2012). A Review on Theoretical Consideration and Types of Models in Hydrology. *J. Environ. Sci. Tech.*, **5**, 249–261.

Koschke L., Lorz C., Fürst C., LehmannT. and Makeschin F. (2014). Assessing hydrological and provisioning ecosystem services in a case study in Western Central Brazil. *Ecol. Processes*, **3**:2, 1–15.

Koschke L., Fürst C., Lorenz M., Witt A., Frank S. and Makeschin F. (2013). The integration of crop rotation and tillage practices in the assessment of ecosystem services provision at the regional scale. *Ecol. Ind.*, **32**, 157–171.

Koschke L., Fürst C., Frank S. and Makeschin F. (2012). A multi-criteria approach for an integrated land-cover-based assessment of ecosystem services provision to support landscape planning. *Ecol. Ind.*, **21**, 54–66.

Kühl A. M., Cardoso da Rocha C. L. M., Espíndola E. L. G. and Lansac-Tôha F. A. (2010). Rural and Urban Streams: Anthropogenic Influences and Impacts on Water and Sediment Quality. *Intern. Rev. Hydrobiol.*, **95** (3), 260–272.

Lima J. E. F. W., Sana E. E., da Silva E. M., Oliveira E. C. (2004). Levantamento da area irrigada e estimative do consum de água por pivôs-centrais no Distrito Federal em 2002. Anais do III Simpósio de Recursos do Centro-Oeste: 1–8.

Lorz C., Neumann C., Bakker F., Pietzsch K., Weiß H. and Makeschin F. (2013). A web-based planning support tool for sediment management in a meso-scale river basin in Western Central Brazil. *J. Environ. Man.*, **127**, S15–S23.

Lorz C., Abbt-Braun G., Bakker F., Borges P., Börnick H., Fortes L., Frimmel F. H., Gaffron A., Hebben N., Höfer R., Makeschin F., Neder K., Roig H. L., Steiniger B., Strauch M., Walde D. H., Weiß H., Worch E. and Wummel J. (2011). Challenges of an integrated water resource management for the Distrito Federal, Western Central Brazil: climate, land-use and water resources. *Environ. Earth Sci.*, **65**, 1575–1586.

Menezes J. M., Prado R. B., da Silva G. C., Mansur K. L. and de Oliveira E. D. (2009). Water quality and its spatial relation with sources of natural and human contamination: Sao Domingos river basin, Rio de Janeiro state, Brazil. *Engenharia Agric.*, **29** (4), 687–698.

Mu Q., Zhao M. and Running S. W. (2011). Improvements to a MODIS global terrestrial evapotranspiration algorithm. *Remote Sensing Environ.*, **115**, 1781–1800.

Myneni R. B., Hoffman S., Knyazikhin Y., Privette J. L., Glassy J., Tian Y., Wang Y., Song X., Zhang Y., Smith G. R., Lotsch A., Friedl M., Morisette J. T., Votava P., Nemani R. R. and Running S. W. (2002). Global products of vegetation leaf area and fraction absorbed PAR from year one of MODIS data. *Remote Sensing Environ.*, **83**, 214–231.

Neitsch S., Arnold J. G., Kiniry J. R. and Williams J. (2011). Soil & Water Assessment Tool Theoretical Documentation Version 2009. Texas Water Resources Institute Technical Report No. 406, College Station, Texas, p. 618.

Oliveira L. and Talamini E. (2010). Water resources management in the Brazilian agricultural irrigation. *J. Ecol. Nat. Environ.*, **2**(7), 123–133.

Orlikowski D., Bugey A., Périllon C., Julich S., Guégain C., Soyeux E. and Matzinger A. (2011). Development of a GIS method to localize critical source areas of diffuse nitrate pollution. *Water, Sci. & Technol.*, **64** (4), 892–898.

Pattanayak S. K. and Wendland K. J. (2007). Nature's care: diarrhea, watershed protection, and biodiversity conservation in Flores, Indonesia. *Biodiv. Conserv.*, **16**(10), 2801–2819.

PGIRH (2006). Plano de gerenciamento integrado de recursos hídricos do Distrito Federal PGIRH/DF 1–6., GDF Brasília, Secretaria de Infra-Estrutura e Obras.

Rounsevell M. D. A., Pedroli B., Erb K. -H., Gramberger M., Busck A. G., Haberl H., Kristensen S., Kuemmerle T., Lavorel S., Lindner M., Lotze-Campen H., Metzger M. J., Murray-Rust D., Popp A., Pérez-Soba M., Reenberg A., Vadineanu A., Verburg P. H. and Wolfslehner B. (2012). Challenges for land system science. *Land Use Pol.*, **29**, 899–910.

Schmidt N., Geier P., Mannschatz T. and Matschullat J. (2009). The cerrado biome in central Brazil - natural ecology and threats to its diversity. *Forum Geoökologie*, **20**(2), 44–50.

Singh V. P. and Frevert D. K. (2006). Watershed Models. Taylor and Francis Group, USA, p. 653.

Strauch M. and Volk M. (2013). SWAT plant growth modification for improved modeling of perennial vegetation in the tropics. *Ecol. Model.*, **269**, 98–112.

Strauch M., Bernhofer C., Koide S., Volk M., Lorz C. and Makeschin F. (2012). Using precipitation data ensemble for uncertainty analysis in SWAT streamflow simulation. *J. Hydrol.*, **414–415**, 413–424.

Strauch M., Lima J. E. F. W., Volk M., Lorz C. and Makeschin F. (2013). The impact of best management practices on simulated streamflow and sediment load in a central brazilian catchment. *J. Environ. Manag.*, **127**, S24–S36.

Zeilhofer P., Rondon Lima E. B., Rosa Lima G. A. (2006). Land use effects on water quality in the urban agglomeration of Ciuabá and Várzea Grande, Mato Grosso State, central Brazil. *Urban Wat. J.*, **7**(3), 173–186.

Chapter 5

Urban structure types and their impact on water resources: A case study in the Distrito Federal of Brazil

R. Höfer, F. Bakker, N. Günther, L. Firmbach, H. Roig, C. Lorz and H. Weiss

5.1 INTRODUCTION

The rapid urban growth of the recent decades concerns mainly developing and emerging countries. Urbanization, that is, surface sealing, higher water consumption and waste water production, is likely to have substantial effects on water resources, for example, shift in infiltration-runoff ratio, reduced groundwater recharge, waste water collection (treatment and drainage) and water quality issues. The analysis and prediction of the effects of urbanization are a major challenge for applied research.

Within the IWAS-ÁGUA DF project (see chapter 1 of this book), the analysis of the effects of urbanization was a key issue (Lorz *et al.* 2012). For the monitoring of urban areas using remote sensing techniques is a cost and time efficient alternative compared to traditional surveys (Taubenböck & Esch, 2010). To characterize the very heterogeneous urban area, an approach for generalization is necessary. The concept of Urban Structure Types (UST) can be applied to characterize the urban area and to represent water-relevant information. The implementation of UST in the study area, the DF of Brazil, was the focus of the project group consisting of scientists from the Helmholtz-Center for Environmental Research (UFZ) and the following Brazilian Partners: CASEB (Companhia de Saneamento Ambiental do Distrito Federal) and UnB (Universidade de Brasília).

The *Distrito Federal do Brasil* (DF), originally planned for approximately 600,000 inhabitants (Paviani, 2002), is situated in the tropical savannas with pronounced dry and rainy seasons. Since it was founded in 1960, the population has increased to more than 2.5 million in 2010, with a population density higher than 440 inhab./km (IBGE, 2012). The population growth concentrates on the satellite towns which enclose the capital of Brazil. For the future, a substantial population growth caused by inner Brazilian migration is expected for the region with increasing negative effects on the local water resources. For the DF, a population of more than 3.0 million is predicted for 2020 (IBGE, 2012).

5.1.1 Background and objectives

Population growth and land use/land cover (LULC) changes caused by urbanization are very dynamic processes, which will maintain their fast dynamics also in the future. On a medium and long-term point of view, climate change will add to the problem with a slight decrease of annual rainfall due to a prolonged dry season and with heavy rain events (see Chapter 2 of this book). The DF and the surrounding region form one of the large urban agglomerations of Brazil that runs into the constraints of system capacities for water supply. In addition, the demand for water will even increase further in the near future (ANA, 2009; CAESB, 2010). The region's local water resources are expected to be increasingly affected by the substantial population growth. These impacts are manifold and can be observed in terms of quality and quantity of the water resources. They include planned and unplanned urban development, uncontrolled use of ground water, water contamination, the urban drainage system, conflicts between urban water consumption and erosion caused by construction activities (Leite da Silva *et al.* 2007). An estimation of the effects of urbanization on the hydrological cycle based on the available data goes along with various difficulties. The recent census (2010) provides a great number of water-relevant information. However, the temporal resolution could not keep pace with the fast-growing urban area.

The overall objective of this study is to assess the impact of UST on water resources in the DF. The following research questions can be formulated, (1) How can UST help to monitor and represent urban areas? (2) Which water-relevant parameters can be represented by UST?

5.2 STATE OF THE ART

Monitoring and characterization of the urban area are of importance when analysing the impacts on the water cycle. This is especially true for areas which are characterized by high dynamics of urban growth where a great part of maps and other conventional sources of information are outdated within a short time. So far, there are different possibilities to characterize urban areas in terms of water-related issues. However, most of the existing studies tend to focus only on one or two specific aspects (e.g., Douglas, 1985; Ellis & Revitt, 2008).

5.2.1 General aspects of water-related issues in urban areas

Anthropogenic impacts by different land-use activities are major factors in the interference with runoff and water quality. The pollution risk in urban areas can emanate from point sources but also from sewage systems and rainwater infiltration, which makes the identification of pollution sources and their impact on urban surface and groundwater difficult (e.g., Vázquez-Suñé *et al.* 2005; Strauch *et al.* 2008; Ellis & Revitt, 2008; Ellis *et al.* 2012). Increasing surface runoff, sediments and contaminants, for example, by nutrients from suburban and urban lawns and gardens are to a large extent caused by an increasing area of impervious surfaces (Carlson, 2007). During dry season, pollutants are deposited on these surfaces and are washed off during rainy season which results in higher pollution loads (Melesse & Wang, 2007).

Domene and Saurí (2006) analysed the factors impacting water resources caused by urban population, for example, demographic, behavioural and housing factors. Most important for explaining the varying water consumption within a city are factors like income, housing type, (number of) household members, the presence of outdoor uses, plant species and consumer behaviour. The study also shows that water consumption is more dependent on the housing type (single family houses vs. multi-storey apartments) than on income classes, since a very high amount of water is used to irrigate backyards (in summer even up to 50%) (Vidal *et al.* 2011).

In urban areas, water pathways have been changed substantially. The contribution of infiltration is reduced due to an increase of direct recharge by leaking water mains and septic tanks. However, the general groundwater recharge in urban areas is regarded to be as high or higher than in equivalent rural areas (Lerner, 1997; Vázquez-Suñé *et al.* 2005). Complex city structures and the change of temporal variability make it difficult for planners to quantify the urban recharge (Lerner, 2002). Mohrlok *et al.* (2008) tried to quantify infiltration processes in urban areas. Taking the lack of data, the large variability and the strong non-linearity of the most important parameters as a starting point, they developed an approach of simplified descriptions of the physical processes determining groundwater recharge and contaminant transport in urban areas.

Ellis *et al.* (2012) stated that data at an optimum spatial and temporal resolution is, in general, rarely available. However, remote sensing data might be applied in several domains (Meijerink & Mannaerts, 2000).

5.2.2 The concept of Urban Structure Types

Apart from existing analyses, which tends to focus on one or two specific aspects, the UST concept provides the opportunity to estimate the impact of human activity on the hydrological cycle. The advantages of UST are the following: (1) UST can be measured directly; (2) their derivation is cost- and time-efficient and (3) the approach is transferable to other situations. Thus, UST might contribute to the assessment of qualitative and quantitative effects on the water balance (Höfer *et al.* 2011).

Banzhaf and Höfer (2008) provide a general definition of UST after Wickop *et al.* (1998): *'Urban Structure Types are spatial indicators that help to divide and differentiate the urban fabric into open and green spaces, infrastructure and building complexes so that their typical characteristics such as physical, functional, and energetic factors can be identified'.*

The present study assumes that these physical, functional and energetic factors are symptomatic of urban structures and can be related to water-relevant parameters. The UST were distinguished according to their physical and functional characteristics as well as depending on the research objective. The characterization of UST includes information about building structure, amount of green area, impervious surface, population density, land use, etc. The classification keys for the case study area were developed during various research stays from 2009 to 2011. Different water relevant parameters can be linked to UST. These parameters based on information obtained from (a) remote sensing data and (b) from the recent census (2010) (IBGE, 2012) are shown in Table 5.1. The variables correspond to water-related processes such as infiltration/runoff, erosion potential (depending on the amount of vegetation, impervious surface and bare soil) and water consumption (CAESB, 2010). These water-relevant characteristics of the UST might be linked to parameters from the outflow of waste water treatment plants

(WWTP). In respect of urban water infrastructure UST represent population density, household density, connection to the sewer system, or the use of septic tanks. These parameters help to calculate corresponding waste water parameters like connection rate or the population equivalent (ATV, 2000). The variable 'water consumption' (derived from UST) allows to deduce values for waste water and effluent water and to identify possible misconnections.

Table 5.1 Overview about water-relevant variables selected from census and RS data.

Variable	Description
From census data (based on setores censitários)	
pop_dens	Population density calculated in inhabitants per square kilometer
hh_dens	Household density as average number of households per square kilometer
inh_hh	Average number of persons per household
bath	Average number of bathrooms or sanitary per household
ws_g	Average number of households with water supply by the general system on the property
ws_ws	Average number of households with no or other forms of water supply (well, spring)
sd_g	Average number of households connected to the general sewer system
sd_st	Average number of households using septic tanks for sewage disposal
sd_rc	Average number of households using rudimentary cesspits for sewage disposal
sd_no	Average number of households without sewage disposal (direct discharge into rivers or lakes)
wd_g	Average number of households using a waste disposal service
wd_no	Average number of households without waste disposal service (waste is burned or buried on the property, disposed on derelict land or into rivers or lakes)
From remote sensing data (based on quadras)	
veg	Average amount of vegetation
imp	Average amount of impervious surface
roof	Average amount of roof area
bare_soil	Average amount of bare soil
water	Average amount of water area

5.3 METHODOLOGY

Three different case study municipalities were chosen, Planaltina in the northeast, and Ceilândia and Taguatinga in the west of the DF (Figure 5.1). These three satellite towns are characterized by enormous demographic changes and changes in LULC during the last years. The population density in all of the case study areas is higher than 2500 inhab./km^2 (IBGE, 2012).

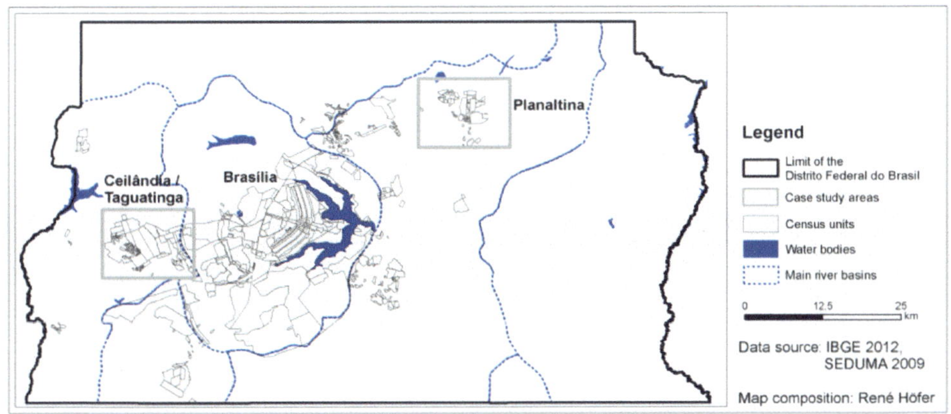

Figure 5.1 Overview about the case study areas.

The wider area of Planaltina is one of the oldest inhabitated areas in the DF. Planaltina was founded in 1859, and in 1960, it was incorporated into the DF. Taguatinga was founded in 1958. It was originally planned for 25,000 inhabitants and supposed to provide housing for the domestic workers in Brasília. Ceilândia was founded in the 1970s by the Brazilian government in the course of the national campaign to eradicate favelas (*Campanha de Erradicação de Favelas*). Since the 1960, the population growth has been driven by relocation processes from Brasília to the satellite towns as well as by internal and external migration processes and a high natural population development. By 2010, the number of inhabitants in the satellite towns had grown significantly: Planaltina had approx. 170,000 inhabitants, Taguatinga 360,000 and Ceilândia 402,000 inhabitants. The building structures of the three study areas are quite different: While Planaltina and *Ceilândia* are characterized by detached and non-detached houses mainly (approx. 95%), only 75% of the buildings in Taguatinga are of this type, 24% are apartment blocks.

Almost every household in the three study areas is connected to the water supply system. Taguatinga is completely connected to the general sewer system, while in Planaltina and Ceilândia only 79 to 83% of the households connect to the general sewer system (others use septic tanks or rudimentary cesspits). In Ceilândia, approx. 20% do not use the public waste disposal while Planaltina and Taguatinga are completely connected. The population structures show similar distributions for Planaltina and Ceilândia (23% of the people are aged below 14, 65% between 15 and 59 and 10% are older than 60 years). In Taguatinga, the number of elderly (18%) grew at the expense of the population below 14 (16%). In Planaltina and Ceilândia, almost three quarters of the households have only up to 5 minimum wages available. In Taguatina, the number of households with only up to 5 minimum wages is less than 45% (for details see PDAD Ceilândia (2011), PDAD Planaltina (2011) and PDAD Taguatinga (2011)).

In general, remote sensing data can be applied in water management in several domains like survey and mapping, spatial analysis and forecasting and decision making (Meijerink & Mannaerts, 2000). Due to the developments in satellite remote sensing with geometric resolutions of less than 1 m, remote sensing data can be used to differentiate urban surfaces such as impervious surface, roofs and road network (e.g., Gamba & Dell'Acqua, 2007; Ehlers, 2007; Weng, 2012). Very high resolution remote sensing data (VHR RS) (Quickbird image with 0.6 m (panchromatic) and 2.44 m (multi spectral) from August/September 2008) was acquired for the case study areas. The Quickbird images were pan-sharpened to be able to use the high geometric resolution of the panchromatic band together with the high spectral resolution of the multi-spectral band. Further input data used for the analysis of socio-economic and socio-demographic information are recent census data and GIS data, for example, administrative limits. The pre-processed image in combination with GIS data is used to characterize the urban area with the help of UST. The different data sources were analysed conjointly by using a semi-automatic OBIA (Object Based Image Analysis) approach. OBIA consists of three main procedure steps. Firstly, the segmentation which divides the satellite image into homogeneous objects. Next, the development of a class hierarchy and a class description with the objective to describe basic land use/land cover classes as clearly and explicitly as possible. The final step is the classification and validation process (for details on OBIA see Blaschke, 2010).

Figure 5.2 gives an overview about the analysis of the study areas. First of all, an UST classification key was developed for the study area. The existing UST for the DF were differentiated using categories such as building structure, amount of green area, impervious surface, population density, land use, etc. On the first level, the classified UST can be distinguished into open spaces, residential areas, public areas, commercial areas and industrial areas. On the next two levels, the UST are further divided. For example, residential areas are divided into detached and semi-detached houses, row houses and apartment blocks. Finally, on the last level, the residential areas are again further divided according to building density, building size, lot size, etc. (Table 5.2).

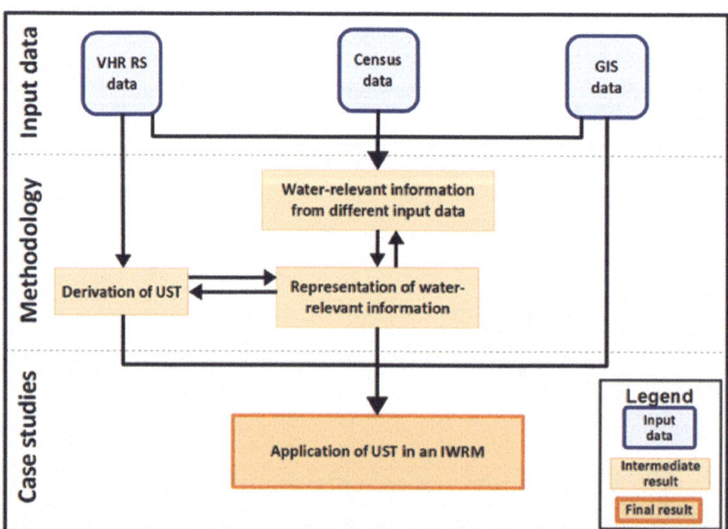

Figure 5.2 Flowchart of the methodology applied in the case study areas.

Table 5.2 Urban structure types classification key (selection) (images quickbird satellite and own photographs).

UST	Description	Visual example	UST detail
P – Parks and green spaces, recreation areas, conservation units	Areas with a low degree of impervious surface and a high amount of vegetation. Mostly public areas in neighborhoods, sport facilities and green areas within transportation infrastructure.		
RH1 – Residential houses, very low density	Areas with small detached houses of rural character, mostly one story high. Large lot sizes. Low amount of impervious surface. No connection to the urban water infrastructure.		
RH2 – Residential houses, low density	Detached houses with a floor space of up to 100 m². Large lot size (up to 800 m²). Low amount of impervious surface. Some areas of this type are connected to the public water supply and waste collection. Most of these areas are in an advanced state of illegal construction.		
RH3 – Residential houses in marginal areas	Small houses of poor construction material on small lots (up to 150 m²). Low amount of vegetation and impervious surface. Most of the open spaces are bare soil. The areas are connected to the water supply system but the drainage and sewer system is still under construction. Some of these areas are illegal but in process of regularization.		
RH4 – Residential houses, high density	Homogeneous areas of high urban density. The buildings are mostly one, sometimes two stories high. The latter are located along main roads. Commercial use with small stores on the ground floor is possible. Lot size of up to 250 m². High amount of impervious surface. The areas are connected to the urban water infrastructure.		

RH5 – Residential houses, medium density	Areas with single family houses. The buildings are one or two stories high and of residential use. Lot size of up to 500 m². The amount of impervious surface varies from 50% to 75%. Some lots have swimming pools and small yards. Connection to the urban water infrastructure exists.	 
RH6 – High-standard residential houses	Residential houses of high quality, mostly two stories high. Large lot sizes of up to 1,000 m². Low amount of impervious surface. Several houses have swimming pools and yards. The areas are connected to the urban water infrastructure and use septic tanks and rudimentary cesspits for sewage disposal. The buildings and infra-structure of this UST were constructed illegally.	 
RB1 – Apartment blocks	Apartment blocks of two to six stories. Characterized by high population density, a low to medium amount of vegetation and a high degree of imperviousness. The apartment blocks are fully connected to the urban water infrastructure.	 
PB – Public buildings	Public buildings and service centers like schools, universities, hospitals, churches, and public sports and leisure centers. Varying amount of impervious surface and vegetation. Generally, the areas are fully connected to the urban water infrastructure.	 
I – Industrial areas	Areas with high building density. Large buildings of mixed use (storage buildings/industrial sites). High amount of impervious surface, no vegetation. This UST includes for example, gas stations and repair shops which often cause high pollution rates.	 

UST	Description	Visual example	UST detail
M2 – Mixed commercial/ residential areas	Areas in the residential sector which are also used for commercial purposes. Especially along main roads. The buildings are up to four stories high. Very low amount of vegetation, high imperviousness. These areas are connected to the urban water infrastructure.		
T – Transportation areas	Roads and railways but also bus and train stations or airports. Mostly no vegetation and a high amount of impervious surface. These areas are sources of pollution and show a high runoff potential.		

The UST were classified on the administrative level of *quadras*, that is, spatial units which correspond to street blocks. These quadras are relatively homogeneous in terms of the contained UST. Nevertheless, census information is only available on the lower administrative units of *setores censitários*, which are artificially created units including several street blocks and hence, several UST. Therefore, the results of the classification process were transferred to the same administrative units for which census data were available. Water-relevant information was selected from remote sensing satellite images and census data (Digital Globe, 2008; IBGE, 2012).

On the basis of a literature analysis, water-relevant parameters were selected from different input data. Information from RS data includes for example, the degree of impervious surface, vegetation and bare soil. The parameters derived from the recent census are based on household level and include information on residents (e.g., populations density, number of residents per household), facilities (number of bath rooms per household) and about urban water infrastructure and waste management. The local water supplier provided information about the average water consumption of an exemplary area (CAESB, 2010). All the information was normalized for each census unit and attribute.

The statistics of the census and remote sensing attributes were calculated for each UST. Descriptive statistics was used to analyse the mean, standard deviation and variance of each attribute per UST. The coefficient of variation (CV) was calculated for all UST. The CV is defined by the ratio of the standard variation to the mean (cf. Eq. 5.1). It describes the dispersion of a variable independently of scale and dimension (independently from variable units)

$$CV = \frac{\sigma}{\bar{x}} \tag{5.1}$$

where CV is the coefficient of variation, σ is the standard deviation, $\bar{x}$ is the mean. The higher the CV, the greater is the dispersion (Bruin, 2011).

These results, and therefore the UST, can be integrated into an IWRM in different fields of application, for example, modelling approaches, scenario development or risk analysis.

5.4 RESULTS FOR THE STUDY AREA PLANALTINA

The UST derived from remote sensing data for Planaltina (Table 5.2) using the OBIA approach showed an accuracy of 40 to 70% for the different case study areas. On the first sight, the low accuracies are disappointing. However, a detailed analysis of the error matrix reveals that the low values can be attributed to clear reasons. The results of the UST classification are highly dependent on the quality of the basic LULC classes as well as the exact representation of the object geometry. The high heterogeneity of the urban area and the limitations of the spectral resolution of the Quickbird data as well as the missing height information from for example, LiDAR are the main reasons for the relatively low classification accuracy.

The same factors led to an inaccurate representation of single objects during the segmentation process which results in errors when geometric features are used for the subsequent classification process. Hyper-spectral data and information about building heights could improve the segmentation and classification process. However, these difficulties are not caused by the restrictions of the UST concept in general; it is a methodological problem when deriving UST. Nevertheless, the approach showed a high potential for obtaining fast and objective results. It was possible to achieve an aggregated level of UST which has to be further differentiated using additional information.

In conclusion, it is possible – with some limitations – to classify UST using Quickbird satellite images. Additional information is needed to improve the segmentation process and hence the classification results. The selected water-relevant parameters (Table 5.1) were analysed regarding their representation by UST. Table 5.3 shows the calculated coefficients of variation (CV) for different UST. CV values which are well represented by UST are marked bold. They focus on the main UST in the study area of Planaltina.

Table 5.3 Coefficient of variation of selected water-relevant attributes for UST in the study area of Planaltina. Abbreviations: RH1 – Residential houses with very low density, RH2 – Residential houses low density, RH3 – Residential houses in marginal areas, RH4 and RH4_2 – Residential houses high density, RH5 and RH5_2 – Residential houses medium density, RB1 – Apartment blocks, M2 – Mixed – commercial/residential areas, PB – Public buildings, NA – not available.

Variable	RH1	RH2	RH3	RH4	RH4_2	RH5	RH5_2	RB1	M2	PB
From census data (based on *setores censitários*)										
pop_dens	**0.58**	1.16	**0.28**	**0.36**	**0.27**	**0.28**	**0.35**	**0.53**	**0.43**	**0.48**
hh_dens	**0.61**	1.11	**0.28**	**0.33**	**0.28**	**0.27**	**0.37**	**0.38**	**0.60**	**0.64**
inh_hh	**0.13**	**0.70**	**0.01**	**0.07**	**0.06**	**0.06**	**0.05**	0.78	**0.24**	**0.17**
bath	**0.15**	**0.67**	**0.13**	**0.12**	**0.04**	**0.20**	**0.11**	**0.40**	**0.26**	**0.17**
ws_g	**0.58**	**0.67**	**0.04**	**0.02**	**0.01**	**0.02**	**0.17**	**0.14**	**0.03**	**0.01**
ws_ws	1.47	1.08	2.04	4.47	3.45	2.19	1.55	NA	2.37	2.96
sd_g	1.55	1.27	**0.49**	**0.03**	1.21	**0.04**	1.18	0.07	**0.16**	**0.05**
sd_st	1.30	2.00	0.94	2.62	1.51	1.55	2.01	NA	12.76	1.12
sd_rc	0.89	**0.67**	9.58	1.95	**0.34**	1.25	**0.65**	1.62	NA	NA
sd_no	NA	NA	NA	NA	5.48	NA	NA	NA	NA	NA
wd_g	**0.58**	**0.67**	**0.01**	**0.00**	**0.00**	**0.00**	**0.00**	**0.05**	**0.01**	**0.01**
wd_no	1.15	2.00	1.73	3.65	5.48	2.00	4.07	NA	NA	3.52
From remote sensing data (based on *quadra*s)										
veg	**0.15**	**0.23**	**0.49**	**0.67**	**0.44**	**0.30**	**0.32**	**0.36**	1.34	**0.36**
imp	**0.70**	**0.61**	**0.56**	**0.25**	**0.20**	**0.21**	**0.25**	**0.32**	**0.32**	**0.51**
roof	0.86	1.02	**0.50**	**0.26**	**0.22**	**0.21**	**0.27**	**0.49**	**0.38**	**0.69**
bare_soil	1.13	**0.41**	**0.61**	1.02	**0.38**	0.75	**0.41**	0.95	0.89	0.93
water	NA	NA	NA	NA	NA	NA	NA	NA	NA	NA

The more census units per UST are available, the more robust are the results. For the UST RH6, I and T, it was not possible to calculate any census information. However, the following variables are well represented by UST: population density, household density, persons per household, connection to the general system for water supply and waste collection. Rudimentary cesspits used for sewage disposal are well represented by the UST. In contrast, households without any sewage disposal system or with sewage disposal in septic tanks, and those with water supply by wells or springs are not represented by any of the UST.

From a methodological point of view, the calculation of average values for areas represented by two or more UST can lead to errors. For example, the calculations for Public buildings (PB) posed some problems. Various PB like schools or sports halls consist of large premises with only one or two buildings and large green or open spaces. However, an UST is composed of building structure and open spaces. Assuming for example that the area of a PB makes up for 50% of the total census unit, the proportion of the built-up area is overrepresented and the average value will be rated too high. In case variables show high standard deviations, the values for the given UST will be overdetermined and the calculation of the variables leads to falsely negative results. In contrast, if an area represented by the UST is not completely covered by built-up yet, the calculated value can be too high or low. A poor representation of some census variables can be caused by a small occurrence of a determined UST in the analysed census units as in the case of RH1, RH2 and M2.

The representation of water-relevant information obtained from RS data was performed on the basis of quadras. A high representation for the average amount of vegetation is given for all UST, except for M2. The average amount of impervious surface is well represented by all UST. The amount of roof area is also well represented except for RH2 which is caused by a high standard deviation. The variable bare soil is worst represented by the UST RH1, RH4, RB1, PB and M2. This can be attributed to the low accuracy rate of the basic LULC classification of less than 60%.

Figure 5.3 shows the comparison of three different water-relevant parameters represented (a) by census information and (b) by UST. The maps show that the UST do not have the same level of detail as the census information and are only able to represent average values. On this account, extreme values are not represented. Nevertheless, a high number of water-relevant parameters are represented by the UST. Furthermore, the higher spatial accuracy of UST became obvious. Census units are artificial areas with equally distributed information. However, built-up areas within a census unit are mostly not homogeneous. It can even be the case that census units represent urban and rural areas, which are not covered by built-up at all. UST are characterized by the composition of built-up and open spaces. Although UST units orientate on street blocks, the areas are also chosen artificially. The modifiable areal unit problem (MAUP) needs to be considered (Openshaw, 1984). However, the spatial resolution of UST demonstrates potential to represent the urban built-up area. Further on, UST show a very high temporal resolution compared to census surveys. The temporal resolution of the UST only depends on a cloud free satellite image. Moreover, the identification of UST is cost and time saving in comparison to a census survey. Concluding these findings, UST allow the monitoring and representation of urban areas.

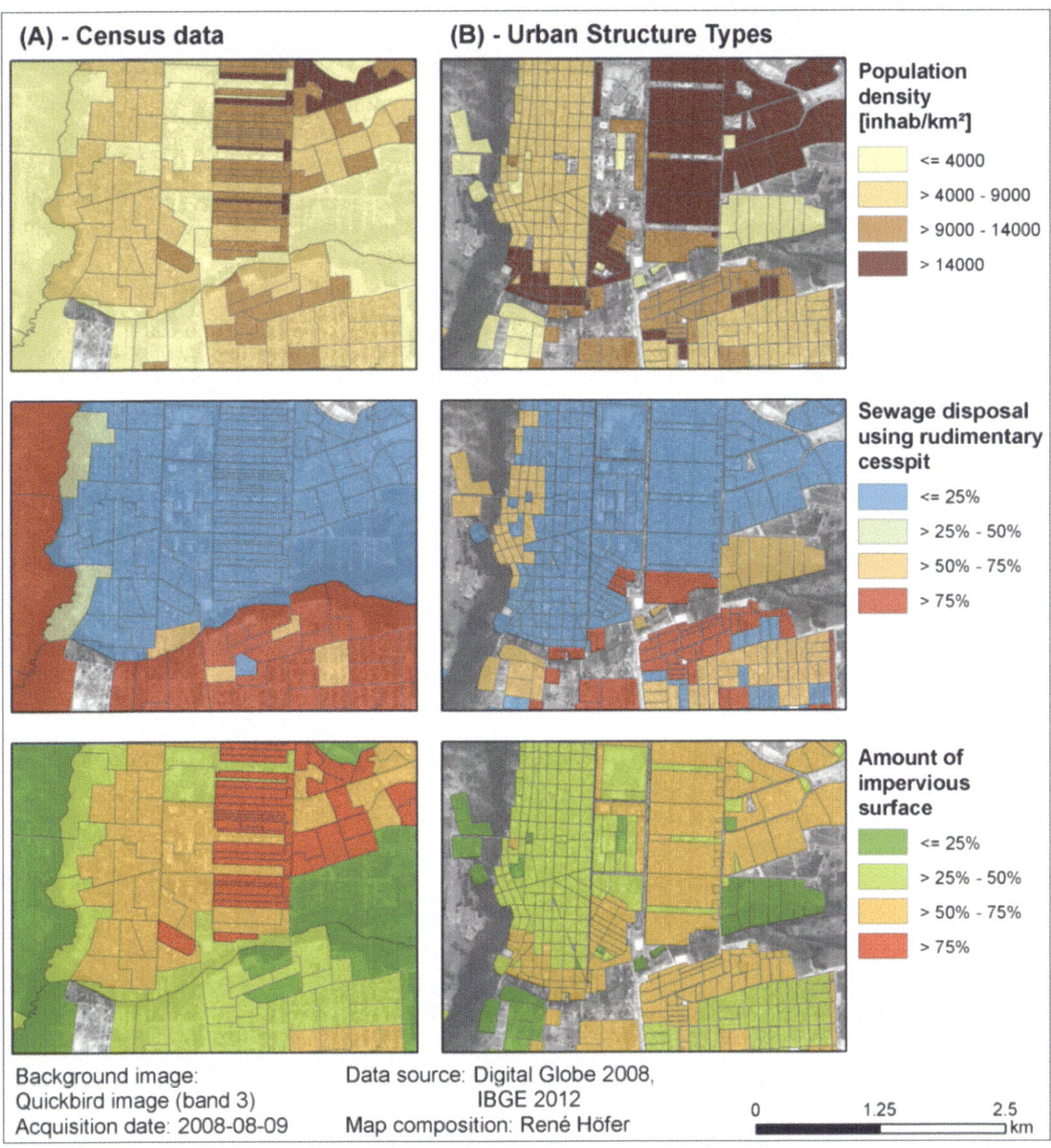

Figure 5.3 Comparison of different water-relevant parameters by census data and UST.

Within an IWRM, the UST can have different fields of application. For example, a risk analysis towards contamination of surface and groundwater pollution caused by urban areas can be performed using UST. The high temporal and spatial resolution leads to a more precise identification of critical areas where to start taking measures from (Höfer, 2013). Concerning the on-going urban growth, a main focus lies on estimating the influence of new urban settlements.

The integration of UST into the planning and decision making process for example, as done in the online planning support tool 'Letsmap do Brasil' (see chapter 4.4), can make use of the UST's potential in predicting the urban impact on the water cycle. According to the high number of water-relevant parameters that are represented by UST, it is possible to include the UST in additional IWRM relevant analysis. Further applications which can be supported by the integration of the UST are SWAT (see chapter 4.3) or groundwater modelling (see chapter 3) as well as sediment analysis (see chapter 4.5). These applications will especially benefit from the high temporal and spatial resolution of the UST.

5.5 CONCLUSION

This chapter has presented the potential of Urban Structure Types for an Integrated Water Resources Management as well as their possibility to spatially represent water-relevant parameters. The main benefit originates from the fact that UST only represent urban built-up area. Further benefits are the higher temporal resolution and the fast and cost efficient processing. However, UST cannot represent the same level of detail as census data. Information about the average water consumption, the amount of waste water and the connection to the sewer system can be represented by UST. Moreover, UST and further RS data allow to derive parameters concerning infiltration capacity, surface runoff and sediment and pollution loads.

Further information about the water infrastructure may help to improve the description of water-relevant parameters by UST. A field campaign on selected UST could verify the parameters of infiltration rate, surface runoff and sediment load and this way improve the risk analysis. UST can be used to monitor urban areas. In particular, they help to obtain up-to-date information about new settlements in highly dynamic areas in a fast and cost-efficient way. The presented approach shows some smaller limitations: the dependence of the results on the classification accuracy of the input UST, the uncertainty in the estimated parameters with a high standard deviation and the spatial differentiation of the UST and built-up areas.

A further improvement of the classification approach may be achieved by including further satellite and LiDAR data and by automating the process. Census data with a higher spatial resolution and field measurements are to be done to improve the accuracy of the represented parameters and/or to find further water-relevant parameters represented by the UST. Finally, the approach needs to be tested outside of the DF.

5.6 REFERENCES

ANA (AgênciaNacional de Águas) (2009). Atlas regiões metropolitanas: Abastecimento Urbano De Água: Resumo Executivo. (Atlas metropolitan region: urban water supply: executive summary), Technical report, Agência Nacional de Águas, Brasília.

ATV-DVWK Deutsche Vereinigung für Wasserwirtschaft, Abwasser und Abfall e.V. (the German Association for Water, Wastewater and Waste), Gesellschaft zur Förderung der Abwassertechnik e.V. -GFA- (Association for the Promotion of Wastewater Technology), Hennef (Ed) (2000), Arbeitsblatt atv-dvwk-Bemessung von einstufigen Belebungsanlagen (Worksheet: Assessment of one-stage activated sludge plant).

Banzhaf E. and Höfer R. (2008). Monitoring urban structure types as spatial indicators with cir aerial photographs for a more effective urban environmental management. *Journal of Selected Topics in Applied Earth Observations and Remote Sensing*, 1(2), 129–138.

Blaschke T. (2010). Object based image analysis for remote sensing. ISPRS *Journal of Photogrammetry and Remote Sensing*, 65(1), 2–16 ISSN 0924-2716. doi: 10.1016/j. isprsjprs.2009.06.004.

Bruin J. (2011). Ucla: Academic technology services, statistical consulting group, (accessed 06 October 2012).

CAESB (2010). Companhia de Saneamento Ambiental do Distrito Federal (Environmental Sanitation Company of the DF) - internal report (unpublished)

Carlson T. N. (2007). Remote Sensing of Impervious Surfaces, Chapter: Impervious Surface Area and Its Effect on Water Abundance and Water Quality, pages 353–368.

Digital Globe (2008). Quickbird imagery products – product guide (revision 4.7.3). Technical report.

Domene E. and Saurí D. (2006). Urbanisation and water consumption: influencing factors in the metropolitan region of Barcelona. *Urban Studies*, 43, 605–1623.

Douglas I. (1985). Urban sedimentology. *Progress in Physical Geography*, 9(2), 255–280.

Ehlers M. (2007). Urban remote sensing, chapter: New Developments and Trends for Urban Remote Sensing, pages 357–376.

Ellis J. and Revitt D. (2008). Quantifying diffuse pollution sources and loads for environmental quality standards in urban catchments. *Water, Air, & Soil Pollution: Focus*, 8, 577–585.

Ellis J. B., Revitt D. M. and Lundy L. (2012). An impact assessment methodology for urban surface runoff quality following best practice treatment. *Science of The Total Environment*, **416**(0), 172–179.

Gamba P. and Dell'Acqua F. (2007). Urban remote sensing, chapter: Spectral Resolution in the Context of Very High Resolution Urban Remote Sensing, pages 377–391.

Höfer R. (2013). Remote Sensing Based Derivation of Urban Structure Types to Assess Hydro- Meteorological Impacts in Highly Dynamic Urban Agglomerations in Latin America, PhD thesis, University of Freiburg

Höfer R., Roig H., Bakker F. and Weiß H. (2011). The impacts of urban dynamics on water resources in the Distrito Federal do Brasil, In 12th International Specialized Conference on Watershed & River Basin Management, 13–16 September 2011, Recife, Brazil, p. 5.

IBGE (2012). Dados do Censo 2010 publicados no Diário, Instituto Brasileiro de Geografia e Estatística, (accessed 10 May 2012).

Leite da Silva C., da Silva C., da Silva, J., de Souza N., and Chacón Arcaya S. (2007). Emprego de Fotografias Áerease Modelo Digital de Terrenono Mapeamento Geotécnico da Área de Protecão Ambiental do Rio São Bartolomeu-DF (The use of aerial photographs, digital elevetation models and geological mapping in environmental protected area of the São Bartolomeu river), In Anais XIII Simpósio Brasileiro de Sensoriamento Remoto, Florianópolis, Brasil, 21–26 abril 2007, INPE, pages 1353–1360.

Lerner D. (1997). Geochemical Processes, Weathering And Groundwater Recharge In Catchments, Chapter Groundwater Recharge, pages 109–150.

Lerner D. (2002). Identifying and quantifying urban recharge: a review. *Hydrogeology Journal*, **10**(1), 143–152.

Lorz C., Abbt-Braun G., Bakker F., Borges P., Börnick H., Fortes L., Frimmel F., Gaffron A., Hebben N., Höfer R., Makeschin F., Neder K., Roig L., Steiniger B., Strauch M., Walde D., Weiß H., Worch E., and Wummel J. (2012). Challenges of an integrated water resource management for the Distrito Federal, Western Central Brazil: climate, land-use and water resources. *Environmental Earth Sciences*, **65**, 1575–1586.

Meijerink A. and Mannaerts C. (2000). Remote sensing in hydrology and water management, chapter: Introduction to an General Aspects of Water Management with the aid of Remote Sensing, pages 329–356.

Melesse A. and Wang X. (2007). Remote Sensing of Impervious Surfaces, chapter: Impervious Surface Area Dynamics and Storm Runoff Response, pages 369–384.

Mohrlok U., Wolf L., and Klinger J. (2008). Quantification of infiltration processes in urban areas by accounting for spatial parameter variability. *Journal of Soils and Sediments*, **8**, 34–42.

Openshaw S. (1984). The Modifiable Areal Unit Problem. Concepts and Techniques in Modern Geography, pages 38–41.

Paviani A. (2002). â Brasília: metrópole incompleta. Minha Cidade, São Paulo, Vitruvius, 02.024: 3, July 2002. URL http://vitruvius.com.br/revistas/read/minhacidade/02.035/2058.

PDAD Ceilândia (2011). Pesquisa distrital por amostra de domicílio – Ceilândia – PDAD 2010/2011 (National Household Sample Survey), Technical report, Companhia de Planejamento do Distrito Federal – Codeplan.

PDAD Planaltina (2011). Pesquisa distrital por amostra de domicílio – Planaltina – PDAD 2010/2011 (National Household Sample Survey), Technical report, Companhia de Planejamento do Distrito Federal – Codeplan.

PDAD Taguatinga (2011). Pesquisa distrital por amostra de domicílio – Taguatinga – PDAD 2010/2011 (National Household Sample Survey), Technical report, Companhia de Planejamento do Distrito Federal Codeplan.

SEDUMA (2009). Secretaria de Desenvolvimento Urbano e Meio Ambiente do Distrito Federal (Secretariat for Urban Development and Environment of the DF).

Strauch G., Möder M., Wennrich R., Osenbrück K., Gläser, H.-R., Schladitz T., Müler C., Schirmer K., Reinstorf F., and Schirmer M. (2008). Indicators for assessing anthropogenic impact on urban surface and groundwater. *Journal of Soils and Sediments*, **8**, 23–33.

Taubenböck H. and Esch T. (2011). Remote sensing – an effective data source for urban monitoring. Earthzine - Quarter Theme Issue on Urban Monitoring, p. 15.

Vázquez-Suñé E., Sánchez-Vila X., and Carrera J. (2005). Introductory review of specific factors influencing urban groundwater, an emerging branch of hydrogeology, with reference to Barcelona, Spain. *Hydrogeology Journal*, **13**, 522–533.

Vidal M., Domene E., and Sauri D. (2011). Changing geographies of water-related consumption: residential swimming pools in suburban Barcelona. *Area*, **43**(1), 67–75.

Weng Q. (2012). Remote sensing of impervious surfaces in the urban areas: Requirements, methods, and trends. *Remote Sensing of Environment*, **117**, 34–49.

Wickop E., Böhm P., Eitner K., and Breuste J. (1998). Qualitätszielkonzept für Stadtstrukturtypen am Beispiel der Stadt Leipzig (Quality objective conception of Urban Structure Types – an example of the city Leipzig), *UFZ Report* **14**, 156.

Chapter 6

Water quality of tropical reservoirs in a changing world – the case of Lake Paranoá, Brasília, Brazil

G. Abbt-Braun, H. Börnick, C. C. S. Brandão, C. B. G. Cavalcanti,
C. P. Cavalcanti, F. H. Frimmel, M. Majewsky,
B. Steiniger, M. Tröster and E. Worch

6.1 INTRODUCTION

6.1.1 The IWAS approach

As a precondition for the general decision on the use of the artificial Lake Paranoá as raw water resource, as well as the choice of the appropriate water treatment technology, detailed data on the water quality are necessary. Whereas long-term data are available for the conventional quality parameters (e.g., nutrients), the data base in view of specific micropollutants is still insufficient. Therefore, in the IWAS-ÁGUA DF project, particular attention was paid to the occurrence of micropollutants. Additionally, the existing data of the conventional parameters were supplemented by some specific analyses.

Recently, the fate of anthropogenic polar organic compounds in the water cycle has gained worldwide attention. Not all of these compounds, such *et al.* as pharmaceuticals, personal care products, fertilizers or pesticides, are eliminated in wastewater treatment plants (WWTPs). Their impact on the aquatic ecosystem and human health is still under discussion (Fent *et al.* 2006). It seems reasonable to expect manifold ecological effects of these so-called anthropogenic micropollutants, in addition to the broad background of refractory organic matter (ROM) (Frimmel *et al.* 2002) also investigated in the project. The identification and quantification of these effects need a sound analytical characterisation of the relevant water constituents (Frimmel & Abbt-Braun, 2011). Due to the drainage of two big WWTPs directly into the lake, these compounds can be enriched in the water cycle and might not be removed by conventional drinking water treatment (Stackelberg *et al.* 2007). Due to the expected population growth in the DF and rising standard of living, the occurrence of persistent compounds in raw water might become more likely in the future (Frimmel & Müller, 2006; Reemtsma & Jekel, 2006; Richardson, 2009). This has to be taken into consideration in the design of the future drinking water treatment plant at Lake Paranoá (Vasyukova *et al.* 2011). Due to the limited information on the occurrence and use of agrochemicals, pharmaceuticals and other emerging pollutants in DF, a combined approach of gathering and evaluating existing data and additional laboratory analyses was used. In this chapter, the results of this study are summarized.

Recent trends and long-term data evaluation of conventional water quality parameters are shown in the beginning of the following sections. Moreover, the additional analyses of nutrients and heavy metals, as well as sum parameters which may influence the selection of the treatment technology, are presented in Sections 6.2 and 6.3, respectively. The occurrence of polar trace compounds in Lake Paranoá is discussed in Section 6.4.

6.1.2 Lake Paranoá

6.1.2.1 General aspects

The artificial Lake Paranoá was constructed in 1959 together with the erection of the new capital of Brazil, Brasília. The primary objective was power generation as well as the improvement of the microclimate for the inhabitants of the new capital. Right from the impoundment the lake also served as a receiver for the – in those days – untreated wastewater of the new capital. Lake Paranoá is situated in the region of the Brazilian Central Plateau, at an altitude of approximately 1000 m above sea level, and is an example for a tropical lacustrine ecosystem. The lake has a surface area of 38 km^2 and a volume of 4.98 * 10^8 m^3, a maximum depth of 40 m and a mean depth of 13 m. Today the lake is additionally used for swimming, surfing and yachting as well as for fishery. The catchment area of 1034.07 km^2 makes up 18% of the DF and is mostly urban, but it also includes some small-scale agricultural areas (8%) (Caldas *et al.* 1999).

The reservoir has four main tributaries: the streams Bananal and Torto in the north, and Gama and Riacho Fundo in the south. Riacho Fundo is the tributary most influenced by urban development, and shows significant differences in water quality parameters compared with the other tributaries (Altafin *et al.* 1995). The tributaries make up approximately 63% of the inflow into the lake, while the remaining 37% is accounted for by direct precipitation into the lake, groundwater, urban drainage and effluents of the two WWTPs. Six principal compartments can be distinguished in the reservoir: the branches of Riacho Fundo (branch A), Bananal (branch E), Gama (branch B) and Torto (branch D), the Central Region (C), and the area near the dam and effluent (F) (see Figure 6.1). Stratification occurs during the rainy season from October to March whereas a circulation is observed in June and July due to lower air temperature. There exists a tendency towards de-stratification during the peak of the rainy season: this phenomenon is probably caused by the higher water flows, which lead to an increase in the water turbulence combined with a decrease in the lake residence time (Altafin *et al.* 1995).

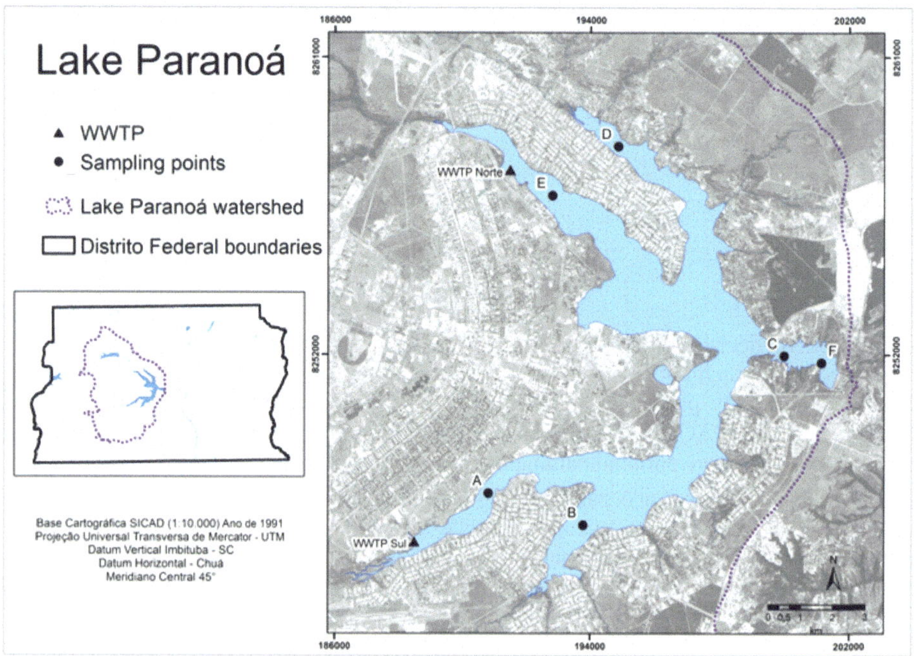

Figure 6.1 Catchment area of Lake Paranoá with sampling points and location of Brasília/DF and the two WWTP South (Sul) and North (Norte) (CAESB/PHII 2013).

6.1.2.2 Water quality: previous studies

Right after its impounding in 1961, Lake Paranoá exhibited excellent physical-chemical and biological characteristics, but then rapidly became eutrophic. In 1962 and 1969, the wastewater treatment plants WWTP South and WWTP North were inaugurated, designed for a population of 225,000 inhabitants and each equipped with secondary treatment. Starting in the 1970s there was a strong deterioration in water quality with massive growth of algae and cyanobacteria (*Microcystis aeruginosa*), as well as large areas of aquatic macrophytes (*Eichhronia crassipes*) and frequent fish kills due to oxygen depletion (Somlyody & Altafin, 1992; Altafin *et al.* 1995; Padovesi-Fonseca & Philomeno, 2004).

In this context, a technical cooperation with the United Nations Development Program (UNDP) was signed and a lake restoration program started (Bjork, 1979; Mattos *et al.* 1992). After a severe bloom of *Microcystis aeruginosa* in 1978, an algae blooming control program was initiated in 1980. Starting in 1978 copper sulfate was applied as an algaecide until 1998 (Altafin *et al.* 1995; Padovesi-Fonseca & Philomeno, 2004). The algaecide was mainly applied to the Riacho Fundo branch (branch A), and to some extent in the central area and the Bananal branch (branch E). Until 1993 more than 92 t were applied. At that time the lake was characterized by a high phytoplankton biomass and a dominance of detritivorous microzooplankton (Padovesi-Fonseca *et al.* 2011). Cronberg (1977) found that approximately 90% of the phytoplanktonic community of Lake Paranoá consisted of the filamentous Cyanophycea *Anabaenopsis raciborskii*, which is now called *Cylindrospermopsis raciborskii* (Branco & Senna, 1991).

A second lake restoration program led to the inauguration of tertiary treatment for WWTP North and WWTP South in 1993. As a result, phosphorus loads in the Riacho Fundo branch dropped by 70%. Nitrogen input was also reduced by 75%. But although there were some improvements in water quality, total phosphorous concentrations remained high (186 µg L^{-1}), chlorophyll-a (100 µg L^{-1}) and transparency did not change significantly (Cavalcanti *et al.* 1997; Grisolia & Starling, 2001). Limnological studies also revealed the maintenance of early conditions, as the dominance of *Cylindrospermopsis raciborskii* and blooms of *Microcystis aeruginosa* (Padovesi-Fonseca & Philomeno, 2004). In 1997, due to a steep drop in temperature during the dry season, an extensive fish kill occurred in the Riacho Fundo branch and approximately 150 t of killed fish were removed. In 1998 a fish count study was conducted, estimating that the overall fishstock in Lake Paranoá exceeded 1400 t and consisted mainly of tilapia and carp (Starling *et al.* 2002). According to Walter and Petrere (2007), this biomass production of more than 300 kg ha^{-1} renders Lake Paranoá amongst the most productive ecosystems when compared to temperate and subtropical lakes.

Despite these efforts, the lake stayed eutrophic until the end of 1998 when a flushing event took place. After increasing the residence time of the reservoir from 321 to 721 days the residence time was reduced abruptly to 192 days in November 1999, removing all the surface water (Angelini *et al.* 2008; Padovesi-Fonseca *et al.* 2009). This led to visible changes in Lake Paranoá: the diversity of zooplankton (Elmoor-Loureiro *et al.* 2004), rotifers (Padovesi-Fonseca *et al.* 2011) and fish (Walter & Petrere, 2007) increased, and the dominant cyanobacteria as well as *Cylindrospermopsis raciborskii* were replaced by green algae and diatoms. Chlorophyll-a values decreased significantly from 50 to 25 µg L^{-1} (Angelini *et al.* 2008) and secchi-depth increased significantly (Padovesi-Fonseca *et al.* 2009). These results suggest that the reservoir was controlled by a community structured under a stable eutrophic state, which was interrupted by the intentional flushing event (Padovesi-Fonseca *et al.* 2009). Recent studies show that the future water quality development of Lake Paranoá is still unclear. Some limnological parameters show a higher trophic state compared to the period of intense eutrophication before 1998 (Padovesi-Fonseca *et al.* 2009).

Monitoring of the water quality of Lake Paranoá by CAESB started in 1976 with an increasing number of parameters and today about 23 physical, chemical and biological parameters are monitored monthly. Within this program heavy metals as well as a set of 21 organic compounds, for example, atrazine, DDT and benzene are also monitored semiannually. For the metals, the limits of quantification (LOQ) allow the comparison to the standards of the Brazilian drinking water legislation [National Council for the Environment, 2005 (CONAMA)]; in the case of organic compounds, they are even lower. None of the values measured have exceeded these standards or the standards set by the WHO (2011).

6.1.2.3 Sediment quality: previous studies

Early studies showed the formation of organic sediment in an area of 0.16 km^2 and a depth of 0.4 m (Bjork, 1979) close the WWTP South. In the other areas of the lake (99% of the bottom area) sediments consisted of the same soil which is typical for the surrounding cerrado *(brownish red)*, with an average concentration of 4.1 mg kg^{-1} phosphorus. In 1987/88 the area around the WWTPs covered with organic sediment of black color had already reached 6 km^2 with a depth of up to 3 m. The average phosphorus concentration was 2.2 g kg^{-1} (Cavalcanti *et al.* 1988). In the rest of the lake the situation resembled the one described by Bjork (1979). During the 1990s, anoxic conditions in the sediment with formation of an anoxic layer, beginning at a water depth of 20 m, were described (Altafin *et al.* 1995).

Three studies deal with the occurrence of different organochlorine pesticides such as DDT or heptachlor in water, sediments and fish from Lake Paranoá. The highest concentrations were found in 1979 (Ministério da Agricultura, 1979). Although banned in Brazil in 1985, Caldas *et al.* (1999) still found DDT in 98% of the fish samples with concentrations up to 77 µg kg^{-1} and in 60% of the sediment samples, but not in water. PCBs, endosulfan, endrin and aldrin were not found above the limit of detection (LOQ). Dianese *et al.* (1976) found higher concentrations of dieldrin in fish (up to 700 times higher than in sediment).

A more detailed study on mineral composition of the sediments was conducted by Maia *et al.* (2005). In 2006, Maia *et al.* studied the distribution of different trace elements in the lake and the tributaries. They found a moderate anthropogenic influence for the reservoir with an elevated geoaccumulation index for Cr, Zn, Ba, Hg and Cu. The Gama and Torto branches showed the least anthropogenic impact. Gioia *et al.* (2006) studied Pb concentration and Pb isotopic composition which allowed for the description of 3 periods in lake history. There was a clear distinction between the period of eutrophication (1970 to 1995) and the recovery of the water quality from 1995 until present. They also found higher Pb concentrations around the two WWTPs, which they correlated to a higher anthropogenic influence (Maruoka & Nascimento, 2006).

6.2 METAL(LOID)S, EUTROPHICATORS AND COLLOIDS

The determination of metals, nutrients and anions is relevant for a comprehensive assessment of the raw water quality of Lake Paranoá and its ecological state. From a toxicological point of view, in particular the concentrations of (heavy) metal(loid)s and their related hazard potential are of concern. They can originate from weathering of the geological native rocks present in the catchment, but are more often related to anthropogenic activities. Moreover, colloids and nanoparticles are of interest, since they can act as pollutants themselves and play a key role in the transport and distribution of sorbed pollutants (e.g., heavy metals, non-polar organics), because of their high surface-to-volume ratio and their high mobility (Delay & Frimmel, 2012, Frimmel & Niessner, 2010). These particles originate from natural and/or anthropogenic sources and are ubiquitous in aquatic environments.

In order to comply with these requirements for water quality assessment, a spectrum of (heavy) metals, metalloids, eutrophying species, anions and particles were selected for monitoring in Lake Paranoá at the sampling sites A to F, in the effluents of WWTP South and WWTP North as previously described as well as in sediments from 2010 to 2013. Analyses were carried out using *optical emission spectrometry with inductively coupled plasma* (ICP-OES), *mass spectrometry with ICP* (ICP-MS) and *ion chromatography* (IC, Tercero *et al.* 2011). Size distributions of particles within the range of 20 nm to 500 nm were measured using *laser-induced breakdown-detection* (LIBD, Bundschuh *et al.* 2001).

With regard to the metals, Al, Mn, Fe, B and Zn could be found in all of the aqueous lake samples from sites A to F, but were clearly below the German and Brazilian threshold values for drinking water (Table 6.1). The data showed no significant temporal and spatial variability. Several metal(loid)s were not detected above the limit of quantification (LOQ) in any of the lake samples, namely As (β <10 µg L^{-1}), Cd (β < 2 µg L^{-1}), Cr (β < 2 µg L^{-1}), Ni (β < 5 µg L^{-1}) and Pb (β < 20 µg L^{-1}). The concentrations in the WWTP effluents were up to 20 times higher (e.g., for Al) than in the lake presenting major point sources of anthropogenic contamination. However, the contribution of these point sources to the total lake pollution requires verification by load calculations.

Table 6.1. Median metal(loid) concentrations c_m measured in aqueous samples of Lake Paranoá.

	Al	As	B	Ca	Cd	Cr	Cu	Fe	Mg	Mn	Na	Ni	Pb	Se	Si	Zn
$\beta_m(x)$ in (µg L^{-1})	34	<10[a]	11	8520	<2	<2	<10	31	994	5	8110	<5	<10[a]	<10	2750	6
Germany (µg L^{-1})	200	10	1000	–	3	50	2000	200	–	50	2*10[5b]	20	10	10	–	–
Brazil (µg L^{-1})	200	10	–	–	5	50	2000	300	–	100	2*10[5b]	70	10	10	–	–

(sampling sites A to F, July 2010 – 2013, grab samples from a depth of 1 m below surface) and drinking water thresholds of Germany (Drinking Water Ordinance, 2011) and of Brazil (CONAMA, 2005); *n* = 70 per metal, *n* = 11 per site; not measured: Sb;a measured by ICP-MS, b indicator value.

A number of metals were measured at higher levels in the tributary Riacho Fundo, which discharges into the same branch of Lake Paranoá as the WWTP South. This particularly holds true for Zn, Sr, Mn, Fe, Al and Ca. Hence, the metals transported by Riacho Fundo are also likely to be a more significant source of contamination as compared to the WWTP emissions. Cu was not detected above the LOQ of 10 µg L^{-1} in any of the lake water samples during the course of the project. Data available from 1979 to 1984 show expectedly that copper concentrations were significantly higher during that period, ranging from 10–200 µg L^{-1} (average = 64 µg L^{-1}, *n* = 60). The results for the other elements are similar to previous data which have been obtained during 2007 to 2010 from grab samples in the different branches once or twice a year. The data showed that the concentrations of the following elements were in general below the quantification limit: Ag (10 µg L^{-1}), As (5 µg L^{-1}), Cd (1 µg L^{-1}), Co (37 µg L^{-1}), Hg (0.4 µg L^{-1}), Li (16 µg L^{-1}), Ni (13 µg L^{-1}), Se (1 µg L^{-1}), Sb (<5 µg L^{-1}), Zn (50 µg L^{-1}). Mean values for Al ranged from 23 µg L^{-1} (branch D) up to 96 µg L^{-1} (branch A). The concentration of Pb was slightly higher in previous studies: 6 µg L^{-1} (branch E) up to 30 µg L^{-1} (branch B).

In terms of nutrients and anions, the concentrations of NO$_3^-$ measured in the WWTP effluent samples (*n = 11 per plant, 24-h flow proportional composite samples*) peaked to 32 mg L^{-1}, but were mainly in the range of 10 to 25 mg L^{-1}. NO$_3^-$ concentrations of grab samples from the tributaries (*n* = 3 per tributary) were below 5 mg L^{-1}. The concentrations in the lake were found to be below 5 mg L^{-1} and hence considerably below the German drinking water threshold of 50 mg L^{-1}. Moreover, the long term data (1976 to present) of the tributaries and the lake indicates that NO$_3^-$ varies to a low extent. This also holds valid for the other anions measured such as Cl$^-$ (*n* = 61, median concentration β_m = 6.6 mg L^{-1},), F$^-$ (β_m = 0.2 mg L^{-1}), NO$_2^-$ (β_m = 0.2 mg L^{-1}) and SO$_4^{2-}$ (β_m = 9.1 mg L^{-1}). Br$^-$ could not be detected in any of the samples (LOQ = 0.5 mg L^{-1}).

In addition to the water quality of Lake Paranoá, the trophic state is of interest since severe eutrophication and algae blooming has occurred in the past (Altafin *et al.* 1995). The content of total phosphorus P_{tot} and dissolved PO$_4^{3-}$ are limiting factors for eutrophication and key indicators to classify the trophic state of stagnant water bodies (Schwoerbel, 1993; Wetzel, 2001). Concentrations of total phosphorus during the last three years were found to range at the limit of quantification of 20 µg L^{-1} (β_m = 22 µg L^{-1}) in aqueous lake samples including the depth profile at sampling site C. Long term data of Lake Paranoá from 1976 to present show the impact of enhanced treatment for phosphorus removal installed at the two WWTPs South and North in 1993 and the consequent regression of P_{tot} and chlorophyll-a (Figure 6.2). According to the established classification schemes based on Carlson (1977), the low values measured of P_{tot}, chlorophyll-a and the

Secchi disk indicate an oligotrophic state. Dissolved ortho-PO_4^{3-}, which is directly available to plants, was not detected in any of the samples above the limit of quantification (LOQ = 0.5 mg L^{-1}). Figure 6.2 also demonstrates that a significant change of the total phosphorus and chlorophyll-a concentrations in the lake due to the reduced WWTP emissions only could be observed after around a decade (ca. 1993 to 2003) due to the inertia of the lake and its resulting respond time. In 1998, dam gates were additionally opened to reduce algae blooms of the lake (Jeppesen *et al.* 2012), which resulted in a drop of chlorophyll-a concentrations from ca. 60 µg L^{-1} to below ca. 10 µg L^{-1} (Figure 6.2). With regard to the data available about the inorganic components, the water of Lake Paranoá meets now the quality standards of drinking water.

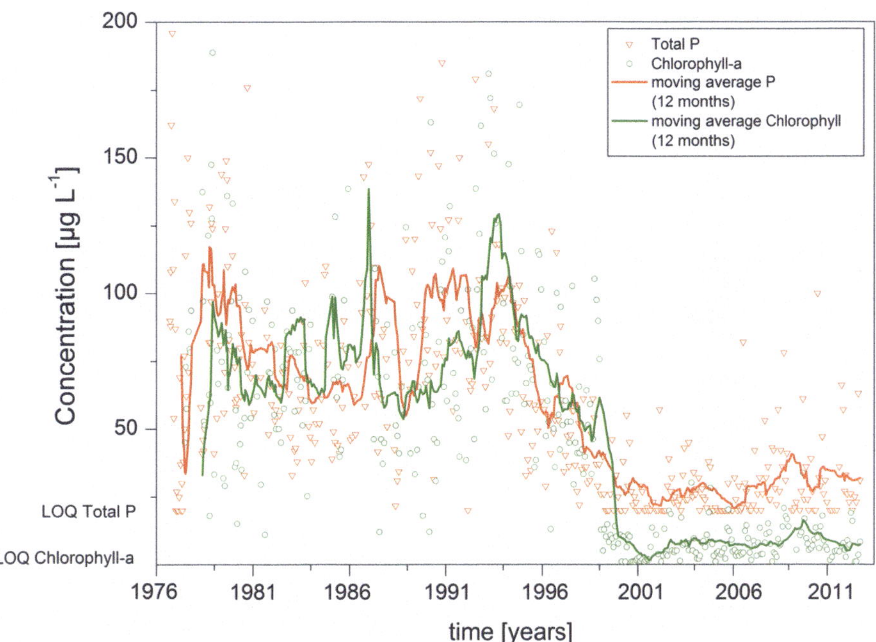

Figure 6.2 Total phosphorus P_{tot} and chlorophyll-a concentrations from 1976 to 2013 at sampling site A nearby the discharge of WWTP South; moving average is the arithmetic mean of the respective previous 12 months; maximum values (>200 µg L^{-1}) not shown; LOQ total P = 20 µg L^{-1}, LOQ chlorophyll-a = 1 µg L^{-1}.

Top layer sediment grab samples from sampling sites A to F (December 2011/January 2012) revealed that metal contents for all metals were found to be higher at the sampling sites in the branches (A, B, D, E) as compared to the middle of the lake at C and F, most likely caused by the loads transported by the tributaries. Highest metal contents were found close to the discharge of the WWTP South. Additional sediment samples nearby WWTP South showed that the sediment contents decreased for all measured metals and anions, with increasing distance from the estuary of Riacho Fundo and the discharge of WWTP South (as sum of Zn, Cr, Ni, Cu, Pb, Cd and As: β_{sum} = 320 mg kg^{-1}) to the sampling site A (β_{sum} = 280 mg kg^{-1}) and C (β_{sum} = 40 mg kg^{-1}) located in the middle of the lake. Caused by the extensive use of $CuSO_4$ as an algicide until the middle of the 1990s (92 t until 1998) mainly in the Bananal and Riacho Fundo branch, copper was expected to be accumulated in lake sediments. However, when compared to the global geogenic background clay rock standard (Müller, 1986; Förstner, 2004), the sediment samples can be considered as being not contaminated with Cu, Cr, Ni and Zn, also not in a depth of 1 m (sample extracted from a drill core). This can be explained by the high sediment loads from the tributaries especially in the Riacho Fundo branch, which built up sediment thicknesses of more than 3 m from 1998 until 2010 according to bathymetric studies (Pires & Ianniruberto, 2008). The present measured Cu contents were found to decrease with increasing distance from the WWTP South discharge (45 mg kg^{-1} near the WWTP effluent, 40 mg kg^{-1} at sampling site A, 6 mg kg^{-1} at C and 17 mg kg^{-1} at F), which may be explained by the punctual use of $CuSO_4$ in the lake areas, which were greatly affected by algae blooms close to the effluents. Cu was detected only in 13% of the WWTP effluent samples (LOQ = 10 µg L^{-1}).

Pb contents were found at 34 mg kg^{-1} at sampling site A and 31 mg kg^{-1} at sampling site E. Accordingly, these sediments are classified as moderately contaminated. The contents of Cd significantly exceeded the global geogenic background at sites A (3.2 mg kg^{-1}), B (2.8 mg kg^{-1}), D (2.6 mg kg^{-1}), E (3.0 mg kg^{-1}) and F (2.1 mg kg^{-1}) and hence, the sediments are classified as moderately-to-heavily contaminated. Only low sediment contents could be observed for total phosphorus (<1.1 mg kg^{-1}). Thus, the heavy phosphorus loads emitted during the 1970s and 1980s could not be detected at least in the top sediment layers.

The potential risk of metal (re-)mobilization from the sediment (n = 17) was assessed by batch tests, with distilled water as solvent (at pH = 6) with a solvent/solid ratio of 10 L kg^{-1}. The highest (re-)mobilization (referred to the total content) was found for Ca with 9 to 12% and Na with 8 to 9%. Sulfur was remobilized up to 11–17%. Cd, Ni and As did not exceed the LOQ (LOQ = 2 µg L^{-1}, 20 µg L^{-1} and 10 µg L^{-1}, respectively) in the eluates of the remobilization experiments. The values for other heavy metals were below 2%. Based on these results, it can be concluded that metal (re-)mobilization from sediments poses a very low risk of contamination.

Colloids were present with number concentrations of up to 1.8 * 10^9 particles per mL^{-1} at a particle size of 20 nm in aqueous samples of Lake Paranoá at sampling point A. The particles exhibited a Pareto-like number based size distribution, that is, strongly increasing particle number concentrations at the lower end of the particle size distribution, resulting in a high surface-to-volume-ratio due to the high numbers of colloids in the lower nanometer range. Based on the assumption of spherical particles, the total particle volume concentrations for the respective sampling sites were in the range between 1.8 * 10^5 µm^3 mL^{-1} (sampling site E) and 6.2 * 10^5 µm^3 mL^{-1} (sampling site A) in March 2012. In September 2012, the concentrations were between 6.7 * 10^4 µm^3 mL^{-1} (sampling site F) and 4.3 *10^5 µm^3 mL^{-1} (sampling site A, Figure 6.3).

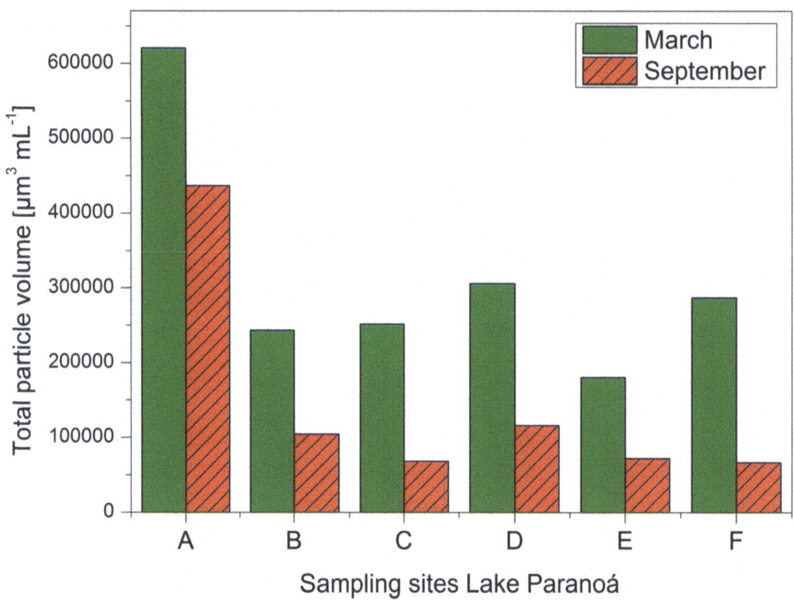

Figure 6.3 Total colloid volume concentrations at sampling points A to F.

The elevated concentrations at sampling point A are likely caused by the tributary Riacho Fundo carrying high sediment loads and the WWTP emissions of WWTP South with total volume concentrations in the effluent of 1.4 * 10^5 µm^3 mL^{-1} to 2.7 * 10^5 µm^3 mL^{-1}. At sampling point E, no elevated concentrations could be observed as compared to the other sampling points indicating that the mass load of the WWTP North effluent is no major pathway for colloid input to Lake Paranoá.

Apart from the mass loads of colloids from the tributaries, atmospheric deposition and diffuse pollution for example, by storm waters, can be additional input pathways. At sampling points C and F, located in the middle of the lake, the colloid concentrations were considerably lower as compared to branch A and in the same range as the other lake branches, indicating that these sampling points are less affected by the tributaries and WWTP effluents, as well as surface water run-off from the lake's banks and the fact that there is no colloid enrichment in the middle of the lake.

Temporal variations of the total colloid content could be observed by comparing the concentrations of the sampling points at different dates of sampling. During the rainy season in March 2012, the concentrations were 1.4 to 4.3 times higher than in October, which might be caused by higher amounts of rainfall. A depth profile showed highest colloid concentrations in depths of 1 m to 5 m. Depending on the density, colloids exhibit no or only slow sedimentation rates and may enrich below the lake's surface, but then decrease with increasing depth. Considering the spatial distribution of the colloids within the lake, the lowest influence of colloids as pollutants or carriers of pollutants on the water quality can be spotted in the middle of the lake, at a depth of 15 to 20 m below the surface.

6.3 CHARACTERIZATION OF DOC AND TOC

Dissolved organic matter (DOM) is taken as an integrated pool of dissolved organic compounds in aquatic systems (Thurman, 1985; Frimmel *et al.* 2002). DOM originating from natural sources is also often called NOM (natural organic matter) or humic substances (HS) (Frimmel & Abbt-Braun, 2011). It plays a dominant role in the biosphere as well as in the treatment of fresh water for human consumption. DOM concentrations and composition can reveal important information about the degree of pollution, the trophic status, but also about the efficiency of membrane or flocculation treatment for drinking water production, and about the formation of disinfection byproducts by using chlorine (Suffet & MacCarthy, 1989; Frimmel & Abbt-Braun, 2009). Although DOM is not toxic it gives the water its taste, odour and a slight colour which is normally not accepted by the consumer. Therefore, DOM is a key parameter for drinking water quality. DOM is typically measured as DOC (dissolved organic carbon), whereas the sum of dissolved and particulate organic carbon is referred to as TOC (total organic carbon).

Size exclusion chromatography (SEC) is one of the most widely used methods for the characterization of DOM. The organic matter is fractionated according to size, charge and hydrophobicity. The combination of chromatographic separation with online detection of UV absorption (at $\lambda = 254$ nm) and dissolved organic carbon (DOC) quantification facilitated the characterization of DOM and by this, the understanding of the origin and process-related changes of DOC (Huber & Frimmel, 1991, 1992; Frimmel & Abbt-Braun, 2011).

Water samples of six different sampling points within the lake (1 m depth below surface, sampling points A to F) were taken as well as samples from the effluents of the two WWTPs and from the tributaries (see Figure 6.1). Quantification of the dissolved and the total organic carbon (DOC, TOC) was performed by a wet chemical method using a Total Organic Carbon Analyzer 820 (Sievers Instruments, Boulder, USA). In addition, SEC was recorded using a SEC-UV/OC system (Huber & Frimmel, 1991; Frimmel & Abbt-Braun, 2011).

The concentrations of DOC sampled in Lake Paranoá and its tributaries are relatively low and range between 1 to 2 mg L^{-1} (Figure 6.4a). These values are lower than those of grab samples collected in 2003 (5.0 mg L^{-1}) and 2005 (2.7 mg L^{-1}) near sample point E (Brasil, 2004; Assis, 2006). The DOC levels in the lake show a low variation during the period of sampling (Figure 6.4b). The highest fluctuations were observed for the rivers discharging into the lake mainly during the rainy season. This can be explained by run-off from the catchment areas, which exhibit natural characteristics or are used for agricultural purposes. For the catchments Bananal, Torto and Gama more than 75% belong to natural and agricultural sites, whereas the Riacho Fundo catchment shows more than 50% of urban land use. As expected, significantly higher DOC values were found for the two WWTP effluents ranging between 5 and 10 mg L^{-1}. The fraction of DOC on TOC (total organic carbon) is very high with > 90% indicating a fairly small particulate fraction (POC). This holds valid not only for the lake samples in 1 m depth, where a lower amount of organic particles could be expected, but also for the WWTP effluent samples as well as for all samples of the depth profile at sampling site C. No significant differences of the DOC concentrations could be observed with increasing depths (1 to 30 m).

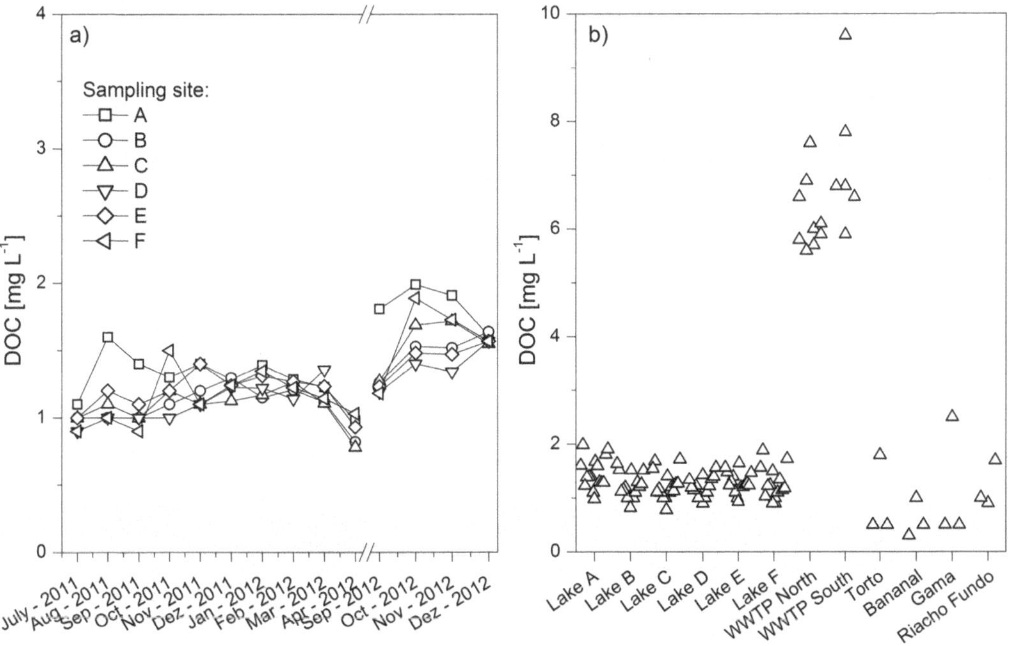

Figure 6.4 DOC concentrations measured at the sampling sites A to F as time series (a) and as comparison between the lake sampling sites ($n = 15$ per site), WWTP effluents ($n = 6$ WWTP South, $n = 9$ WWTP North) and tributaries ($n = 3$) from 2010 to 2013 (b).

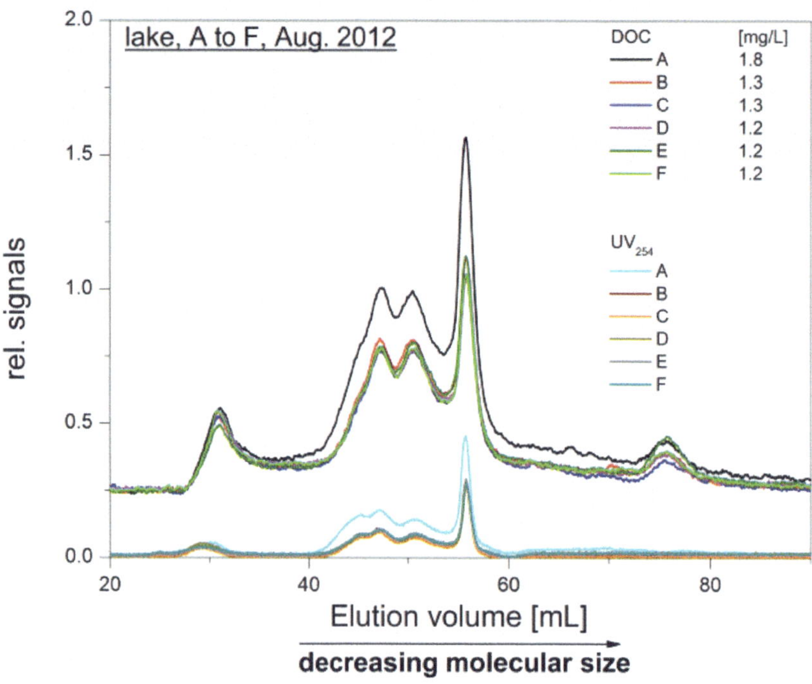

Figure 6.5 Size exclusion chromatograms (organic carbon detection [DOC] and UV [254 nm] detection) of the sampling sites A to F in Lake Paranoá.

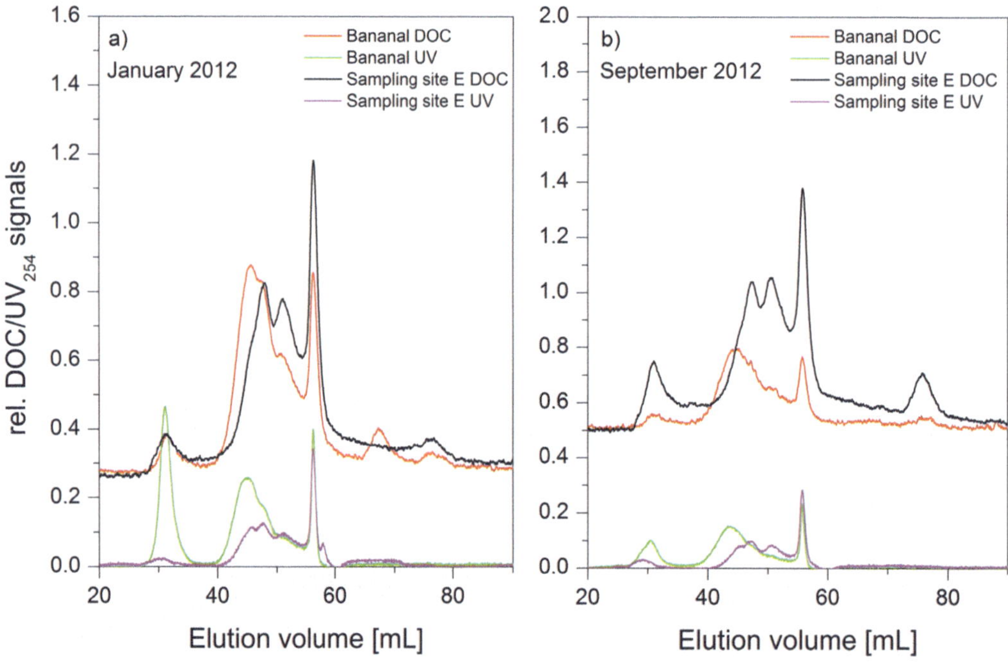

Figure 6.6 Size exclusion chromatograms (organic carbon detection [DOC] and UV [254 nm] detection) of the tributary Bananal and branch E of the lake during rainy season in January 2012 (a) and dry season in September 2012 (b).

The SECs (Figure 6.5) were calibrated by polyethylene glycols and show several nominal size fractions ranging from 20.000 to 60 g mol^{-1} (nominal molecular weight calibration). The molecular weight distribution of the OM in the lake is quite similar (A to F), showing almost no variation in size in view of season and sampling site. There is a low amount of high molecular weight substances, whereas the main fraction (nominal: 4000 to 400 g mol^{-1}) is in the medium molecular weight range and shows aromatic and unsaturated character.

The DOC quality of the tributaries is different during rainy and wet season. Due to the surface run-off water from natural and agricultural sites the quality of the DOC discharging into the lake is different. This becomes apparent from the higher molecular weight characteristics of the DOC, most obvious for the catchment site Bananal (Figure 6.6a, b). With respect to the low DOC concentration and the low variation in season, the lake water can be characterized as raw water with good quality. From the size distribution of the DOC compounds it can be expected that most of main DOC fraction is effectively removed by flocculation during drinking water treatment (Frimmel & Abbt-Braun, 2011). The data will be the basis for the further development of the treatment strategies in view of further needs in (drinking) water treatment.

6.4 THE OCCURENCE OF ORGANIC MICROPOLLUTANTS IN LAKE PARANOÁ

6.4.1 General aspects

The monitoring of organic micropollutants has shifted to an essential part in (drinking) water quality assessment due to their potential adverse effects on human health and ecosystems alike (Schwarzenbach *et al.* 2006). A variety of these organic micropollutants exhibits high polarity and high water solubility, which made their analysis challenging in the past. However, the recent advancement of sensitive LC-MS/MS techniques now allow for their determination at a trace level in the ng L^{-1} and lower µg L^{-1} range (Reemtsma & Jekel, 2006).

During the last two decades, numerous measurement campaigns, primarily in Europe and North America, have been carried out that have demonstrated the ubiquitous occurrence of pharmaceutical residues in domestic wastewater and also identified the effluents of wastewater treatment plants (WWTPs) as a major input pathway of micropollutants into the aquatic environment (Alder *et al.* 2010; Majewsky *et al.* 2013; Reemtsma *et al.* 2006). In particular, pharmaceuticals are partially released untransformed from the human body. The existing conventional municipal sewage plants are not suited to completely remove these emerging pollutants or their stable transformation products. Thus, the upgrade of existing WWTPs with an additional tertiary treatment step and other technical measures are under discussion in Europe (Joss *et al.* 2008).

To advance this cause, the European Water Framework Directive (WDF) (European Commission, 2000) sets up environmental objectives to achieve a 'good water status' for all European water bodies by 2015 and establishes a conceptual framework to achieve these objectives. Here, the main goals are to develop new research initiatives on most pressing issues, and to support the policy making community in the implementation of the WDF with the need to identify 'river basin specific pollutants' (Loos *et al.* 2009).

Furthermore, despite systematic scientific investigations, there is still a lack of knowledge about the long-time occurrence and the behavior of polar micropollutants in the environment and in water treatment processes as well as their effects (particularly combined effects of mixtures of anthropogenic trace compounds) on human and/or ecosystems (Sumpter & Johnson, 2005).

Although mainly surface water is used for drinking water production in Brazil, especially from reservoirs, there is only scarce information about the occurrence of organic micropollutants (Demoliner *et al.* 2010; Jardim *et al.* 2012; Kuster *et al.* 2009; Sodré *et al.* 2007; Sodré *et al.* 2010b; Stumpf *et. al.* 1999). Recent publications show relatively high concentrations of pharmaceuticals and pesticides in Brazilian surface and also tap water (Locatelli *et al.* 2010; Sodre *et al.* 2010a, see Table 6.2).

Most research has been carried out in streams in the state of São Paulo and the state of Rio de Janeiro. However, the number of compounds analyzed in Brazil is generally much smaller compared to data available for Europe and North America. When comparing the data, maximum levels of measured concentrations can be very high, mainly when untreated sewage is discharged uncontrolled directly into surface waters. This makes a comparison of surface water concentrations in Brazil with worldwide data difficult. On the other hand, concentrations found in WWTP effluents are similar to worldwide data. Apart from one author (Stumpf *et al.*1999), there is no information about the behavior of organic micropollutants in tropical lakes. Some authors report a significant difference between rainy and dry season, with increased concentrations of certain compounds in rivers by a factor of 20 or more in the dry season as compared to the rainy season (Locatelli *et al.* 2010). Depending on the applied technologies, several compounds were not removed during drinking water treatment (Stackelberg *et al.* 2007) and were consequently detected in drinking water (Eichhorn *et. al.* 2002; Sodré *et al.* 2010a).

Due to the discharge of two WWTP effluents into the Lake Paranoá and the land use structure of Distrito Federal with numerous agricultural and park areas, the occurrence of typical emerging pollutants as well as pesticides can be assumed. In contrast, the appearance of typical chemicals from industry and waste deposits is not expected because of the low level of industrialization in this region. Given the lack of information about the relevance of organic micropollutants in the Lake Paranoá, one of the objectives of the investigations presented was to get an overview of the occurrence of these pollutants.

Table 6.2. Concentrations for selected pharmaceuticals in Brazilian rivers and effluents from WWTPs.

Compound	Year	Maximum concentrations (μg L^{-1})	
		Rivers	WWTP
Acetylsalicylic acid	1996	1.0/21.0[a]	3.0
Bezafibrate	1996	0.2	1.3
Caffeine	2005	41.7	1.3[b]
Clofibric acid	1996	0.1	1.5
Diclofenac	1996	6.0	1.5
Gemfibrozil	1996	n.m.	1.8
Ibuprofen	1996	0.2	3.8
Naproxen	1996	0.2	3.0
Paracetamol	2005	8.3	5.9[b]
Sulfamethoxazole	2010	0.1	n.m.
Trimethoprim	2010	0.5	n.m.

(data from Stumpf *et al.* 1999; Ghiselli, 2006; Raimundo, 2007; Sodré *et al.* 2007; Froehner *et al.* 2009; Locatelli *et al.* 2010)

n.m. – not measured; [a] rainy/dry season; [b] mean concentration

In Brazil, most pharmaceuticals are sold over the counter without prescription, the only exception being antidepressants and antibiotics (Volpato *et al.* 2005). Data about production or consumption are hardly available. For pesticides, IBAMA (Brazilian Institute for the Environment and Renewable Natural Resources) has published values of application rates for the year 2009 (IBAMA, 2010). Since further data about consumption and use of pharmaceuticals, personal care products and pesticides for Brazil are scarce, more than 50 typical micropollutants of different compound groups (see section 6.4.3) were pre-selected for a scrreening according to data about the worldwide relevance of organic micropollutants in surface waters, as well as their usage (Loos *et al.* 2009; Kümmerer, 2010; Richardson, 2009; Ternes, 2001). In one of the few comprehensive transnational studies, 35 relevant emerging pollutants were monitored in more than 100 rivers in 27 European countries (Loos *et al.* 2009). Here, the most frequently appearing and most highly concentrated compounds were 1*H*-benzotriazole (with the highest median concentration of 226 ng L^{-1}), carbamazepine, tolyltriazole, caffeine, and nonylphenoxyacetic acid. Further relevant substances were naproxen, bezafibrate, ibuprofen, gemfibrozil, perfluorooctanoic acid (PFOA), perfluorooctanesulfonate (PFOS), sulfamethoxazole, isoproturon, and diuron. Only few parts of the investigated rivers showed no or very little pollution. Besides others, the results of this study were used for the selection of target compounds within the micropollutant screening in Lake Paranoá and WWTP effluents.

The strategic approaches of the study presented were (1) to get a first overview of the occurrence and distribution of typical organic micropollutants in Lake Paranoá; (2) to identify and define appropriate key compounds for future monitoring; and (3) to give recommendations for suitable technologies for the future wastewater as well as drinking water treatment.

6.4.2 Sampling and experimental details

In each branch and in the center of the lake, one sampling point (A to E) was selected. In addition, sampling point (F) was included at the future intake point of the drinking water treatment plant planned for Lake Paranoá (see Figure 6.1). Sampling sites are in accordance with the sites used by the regional water supplier CAESB for their long-term limnological monitoring. Samples were collected at 1 m depth and additional depth profiles were sampled at 1, 5, 10, 15, 20 and 30 m at the central point (sampling site C) using a bailer. Overall, 9 sampling campaigns including 13 samplings of Lake Paranoá were carried out between 2009 and 2012. To include the main loads introduced into the lake, further samples of the effluents of the two WWTPs (WWTP North and WWTP South) were taken. In October 2012, the four main tributaries were included in a sampling campaign and sampled twice within two weeks. In the first four sampling campaigns, a wider range of organic compounds was investigated. Later, the selection of analyzed compounds was successively adapted and reduced according to the results of the first sampling campaigns as well as information from recent literature. Thus, complete data sets for all compounds and for all sampling campaigns have not been obtained.

The analytical methods used for the analysis of the different organic micropollutants are described in detail elsewhere (Kutschera *et al.* 2009; Nödler *et al.* 2010; Zwiener & Frimmel, 2004). Samples were collected using 500 to 1000 mL amber glass bottles (Duran, Germany), which had been cleaned in the laboratory and rinsed twice with the sample. The bottles were transported cooled in insulated boxes and stored in the dark at 4°C until solid phase extraction, which was carried out not later than 48 h after sampling. Before extraction, a mixture of surrogate standards was added. Wastewater samples were taken hourly as flow-proportional daily composite samples using a refrigerated automatic water sampler (Teledyne Isco, Lincoln, NE, USA) equipped with a peristaltic pump. Sampling took place from 8 AM until 8 AM of the next day.

Sampling of the WWTP effluents was carried out within a short time lag to the sampling dates of the lake. Both WWTPs follow the same treatment scheme, the only difference was an additional equilibration tank at WWTP North, taking in peak flow events during the day and thus allowing a more constant flow. Both treatment plants are equipped with tertiary treatment for phosphorus removal.

6.4.3 Organic micropollutants in Lake Paranoá – analytical results

6.4.3.1 General results and tendencies

First targeted screenings show the occurrence of a relatively high number of organic micropollutants in the Lake Paranoá. However, the concentrations of the selected compounds analyzed in the lake water are lower in comparison to European lakes (Tixier *et al.* 2003), or even below the limit of quantification (LOQ) of the applied analytical methods. The effluent concentrations of some of the compounds, for instance sulfamethoxazole or iopromide, were within the range known for European WWTPs (Carballa *et al.* 2004). Table 6.3 gives an overview of the investigated compound groups measured between 2009 and 2012.

Table 6.3. List of analyzed compound classes and compounds analyzed (selected examples), measured between 2009 and 2012.

Compound class	Compounds analyzed (selected examples)
X-ray contrast media	Amidotricoic acid, iopromide, iopamidol iohexol
Perfluorinated compounds	Perfluorooctanoic acid (PFOA), perfluorooctansulfonate (PFOS)
Cyanotoxins and microcystins	Microcystin-LR, microcystin-RR, cylindrospermopsin
Pharmaceuticals	Carbamazepine, benzafibrate, phenazone, atenolol, diazepam
Sweeteners	Acesulfame, saccharin, sucralose, cyclamate
Alkaloids	Caffeine, cotinine
Corrosion inhibitors	1*H*-Benzotriazole, tolyltriazole
Drug metabolites	Benzoylecgonine
Pesticides, degradation products	Atrazine, diuron, isoproturon, desethylatrazine
Personal care products	Iso-E-super (OTNE), galaxolide (HHCB), triclosan
Polycyclic aromatic hydrocarbons	Phenanthrene, pyrene, benzo[*a*]pyrene

In Table 6.4, the limit of quantification (LOQ) and the relative occurrence between 2009 and 2010 in the lake water and the two WWTP effluents is shown for selected compounds. As indicated in Table 6.4, several compounds could not be detected, neither in the lake nor in the WWTPs, although they are quite well known for their occurrence in surface waters in Europe and North America. Compounds which were not detected are: clofibric acid (lipid regulator), diazepam (anticonvulsant), mecoprop and metazachlor (herbicides), iomeprol (X-ray contrast media) and fluoxetine and sertraline (selective serotonin reuptake inhibitors, antidepressants). These compounds show low to medium concentrations in European and Brazilian surface waters according to several studies (Stumpf *et al.* 1999, Kümmerer, 2010; Sodré *et al.* 2010a).

A relatively low temporal variability in the occurrence of the investigated micropollutants was observed in the lake as well as in the WWTPs with exception of several compounds, for example, caffeine, phenanzone, or tolyltriazole (later only in WWTP effluents). While one would expect a rise in concentrations in Lake Paranoá for the observed compounds during the dry season, this is only obvious with a weak tendency for some compounds (e.g., caffeine). However, up to now, such seasonal trends are not very pronounced during the dry season. It is most obvious for sampling site A since the flow of the treatment plant is rather stable during the year while discharge from the tributaries and the groundwater reaches a minimum flow in the dry season (see Figure 6.1). The concentrations for most of the compounds analyzed in the central area of the Lake Paranoá are below 100 ng L^{-1} (sampling point C, see Table 6.5). An exceptional situation was found for the values of point A, which were strongly influenced by the effluent of the WWTP South. They showed the highest concentrations in the lake for most analytes, especially in the dry season (May to September).

Table 6.4. Selected compounds analyzed, their limits of quantification (LOQ, sample volume of 500 mL, LOQs for WWTP effluent are higher by a factor of five) and their relative number (RN) of positive detection (>LOQ) in Lake Paranoá and in the WWTPs North and South between 2009 and 2010. RN is given for the samples exceeding the LOQ as percentage of the total number of samples n analysed. Sample numbers from Lake Paranoá: n = 12 for sweeteners, PFOA and PFOS, n = 30 for all other compounds, Sample numbers from WWTPs: n = 6 for sweeteners, PFOA and PFOS, n = 10 for all other compounds.

Compound	Class	LOQ (ng L^{-1})	RN (%); Lake Paranoà (A-F, at 1-m depth)	RN (%); WWTPs
1-H-Benzotriazole	Corrosion inhibitor	5	43	100
1-Methylxanthine	Caffeine metabolite	21	10	40
3-Methylxanthine	Caffeine metabolite	28	3	30
Acesulfame	Sweetener	10	100	100
Aspartame	Sweetener	10	0	0
Atenolol	Antihypertensive	4	90	100
Atrazine	Herbicide	2	43	10
Bezafibrate	Lipid regulator	4	7	10
Caffeine	Stimulant	4	100	100
Carbamazepine	Anticonvulsant	2	43	100
Cetirizine	Antihistamine	2	13	100
Citalopram	SSRI/antidepressant	3	50	100
Clarithromycin	Antibiotic	7	10	100
Clofibric acid	Lipid regulator	3	0	0
Cyclamate	Sweetener	10	42	100
Desethylatrazine	Herbicide metabolite	2	60	0
Desisopropylatrazine	Herbicide metabolite	6	7	0
Diazepam	Anticonvulsant/sedative	1	0	0
Diclofenac	Anti-inflammatory	2	7	100
Diuron	Herbicide	3	0	80
Erythromycin	Antibiotic	7	10	20
Fluoxetine	SSRI/antidepressant	16	0	0
Gemfibrozil	Lipid regulator	2	47	100
Ibuprofen	Anti-inflammatory	4	23	90
Iohexol	Iodinated contrast media	21	60	100
Iomeprol	Iodinated contrast media	19	0	0
Iopamidol	Iodinated contrast media	19	47	40
Iopromide	Iodinated contrast media	19	47	80
Isoproturon	Herbicide	3	0	0
Loratidine	Antihistamine	3	0	0
Mecoprop	Herbicide	1	0	0
Metazachlor	Herbicide	2	0	0
Metoprolol	Antihypertensive	4	13	100
Naproxen	Anti-inflammatory	5	0	80
Pantoprotazole	Gastric acid regulator	5	0	20
Paracetamol	Analgesic	4	3	20
Paraxanthine	Caffeine metabolite	3	100	100
PFOA	Fluorinated surfactant	5	100	100
PFOS	Fluorinated surfactant	5	100	0
Phenazone	Analgesic/antipiretic	2	100	100
Primidone	Anticonvulsant	3	3	50
Roxythromycin	Antibiotic	10	0	0
Saccharin	Sweetener	10	25	100
Sertraline	SSRI/antidepressant	10	0	0
Sotalol	Antihypertensive agent	5	17	100
Sucralose	Sweetener	10	100	100
Sulfamethoxazole	Antibiotic	3	100	100
Tetrazepam	Anticonvulsant	2	0	0
Theobromine	Caffeine metabolite	5	93	50
Theophylline	Caffeine metabolite	3	30	90
Tolyltriazole	Corrosion inhibitor	5	83	100
Trimethoprim	Antibiotic	3	13	100

Table 6.5. Maximum concentrations for selected organic micropollutants between 2009 and 2012.

Compound	Maximum concentrations in (ng L^{-1})		
	Lake Paranoá		WWTPs
	Point A (1 m)	Point C (1 m)	Effluent
Acesulfame	265	170	18,100
Atenolol	285	35	2790
Atrazine	8	9	16
Caffeine	60	79	18,000
Clarythromycin	24	<LOQ	2043
Carbamazepine	32	17	11,300
Citalopram	18	10	587
Diclofenac	10	<LOQ	2520
Gemfibrozil	40	8	761
Ibuprofen	24	3	496
Iohexol	955	21	4425
Iopamidol	338	65	5969
Iopromide	755	85	14,550
Ioxithalamine acid	416	135	22,200
PFOA	4	3	34
PFOS	6	10	3
Phenazone	213	72	1780
Saccharin	<LOQ	7	15,300
Sulfamethoxazole	105	26	727
Tolyltriazole	70	15	814

Within Lake Paranoá, the concentrations of most of the investigated analytes significantly decreased with increasing distance from the WWTPs. The concentrations in sites C or F are lower compared to site A or E. It is interesting to note that the decline in concentrations differs for each compound. For this, dilution may play a role but attenuation processes (photolysis, biodegradation) in the lake might also contribute, as it can be deduced from the different ratios of the individual substances (Schindelin & Frimmel, 2000, Doll & Frimmel, 2004; 2005). Figure 6.7 shows two typical examples, (1) iopromide, a synthetic diagnostic agent used in large hospitals and advanced medical institutions located in urban areas, and by this feeding the public sewer system with this ecologically and technically very stable parent compound. (2) caffeine, a partly degradable natural product used by the broad population in urban and rural populated areas as well. By this, iopromide acts as a fairly ideal tracer for urban waste water and its dilution, whereas caffeine reflects the entire population with organized or decentralized waste water systems with natural attenuation.

With the exception of iopromide and caffeine, concentrations at sampling site F, the site of the future raw water extraction, did not exceed 100 ng L^{-1}. These concentrations are below the values of data shown for polluted surface waters in Europe (Richardson, 2009; Tixier *et al.* 2003; Ternes, 2001), and are not yet known to have any ecotoxicological impacts (Fent *et al.* 2006). Higher concentrations can be measured for point A, influenced by the discharge of WWTP South. Here, phenazone and atenolol are the compounds with maximum concentrations above 100 ng L^{-1} (Table 6.5). For point E, a certain influence of WWTP North can also be seen, but to a much lower degree than for point A.

In the effluents of both WWTPs, an increasing concentration from Sunday to Thursday could be seen for most compounds. In addition to pharmaceuticals and contrast media, high concentrations were also found for caffeine, sweeteners and corrosion inhibitors (Table 6.5).

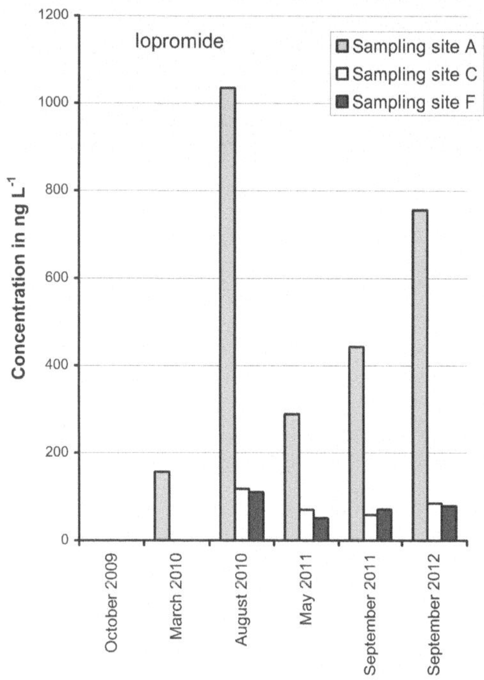
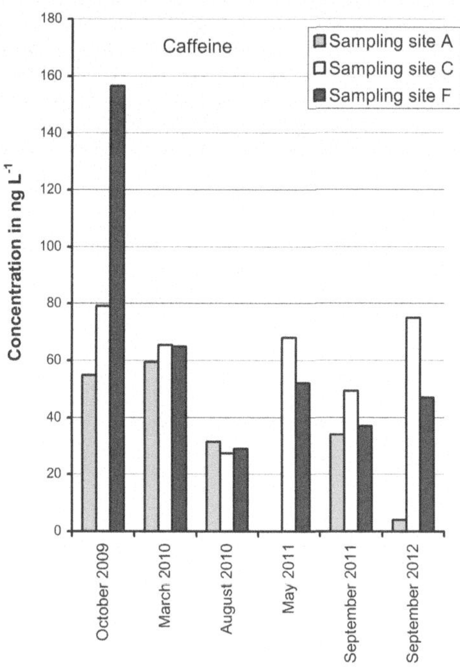

Figure 6.7 Concentration course of iopromide and caffeine at sampling sites A, C and F (1 m below surface), time period from October 2009 to September 2012.

6.4.3.2 Risk matrix and key compounds

Based on the results of the screening, a 'risk of occurrence' matrix was generated for Lake Paranoá. The list is exclusively considering the RN of occurrence and the concentration of the micropollutants. The number of positive results as well as the median concentration levels were clustered to yield three and four values, respectively (Figure 6.8). The risk matrix shows the product of both factors for each compound and is ranking each compound from 0 to 6 according to its occurrence. It has to be noted that the risk matrix does not consider any further effects or a toxicological risk assessment for the compound concerned.

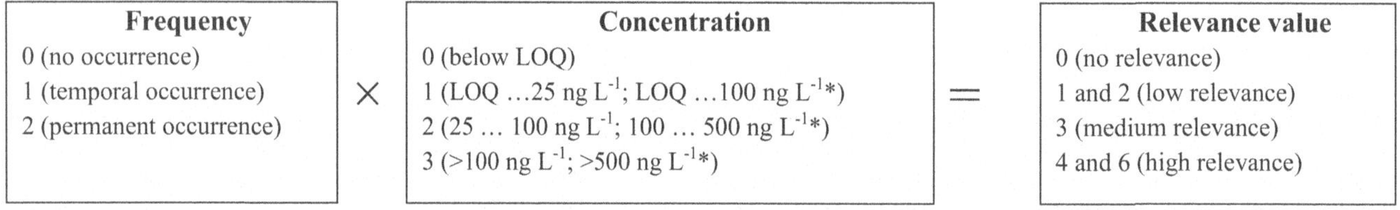

Figure 6.8 Set up of the risk matrix, based on the estimated frequency and concentration of each micropollutant in Lake Paranoá, *WWTP effluents.

The result for sampling sites A and C as well as for WWTP North and WWTP South are shown in Table 6.6. The 'risk matrix' represents a useful tool (1) to improve the future monitoring of water quality; (2) to find out key components which can be used, for instance for the identification of trends; and (3) to serve as decision support to optimize and/or modify the techniques to reach an efficient wastewater as well as drinking water treatment, with the aim of removing micropollutants (Steiniger *et al.* 2011). The 'risk matrix' approach is open for improvement towards the influence of the WWTP discharge and by the inclusion of more compounds and additional parameters (such as persistence/degradability or removability by certain treatment processes) and of course to a more frequent monitoring. This would allow the adaptation of the matrix to different specific requirements, for instance the needs of the local water supplier and the scientific basis for future management concepts. In this context, individual threshold values for each compound can be introduced to reflect different governmental recommendations or ecotoxicological effect levels.

Table 6.6. Selected data from the risk matrix, generated from measured concentrations and RN detected for 50 compounds during the first 5 sampling campaigns for sampling site C in Lake Paranoá and WWTP South.

$n = 50$ compounds	No relevance	Low relevance	Medium relevance	High relevance
Lake Paranoá (Site A)	17	16	9	8
Lake Paranoá (Site C)	31	9	5	5
Effluent (WWTP South)	16	4	14	16
Effluent (WWTP North)	19	3	7	21
Selected compounds	Saxitoxins Microcystins Fluoxetine Aspartame	Carbamazepine Gemifibrozil 1*H*-benzotriazole	Iopromide Iopamidol Sulfamethoxazole	Atenolol Phenazone Caffeine

Due to the predicted population growth in the DF associated with the expected rise in the standard of living, a higher consumption of pharmaceuticals and personal care products can be expected. Longer periods of drought, caused by the climate change, will very likely lead to an increase of the concentrations of pollutants in point C. Therefore, compounds with high and medium relevance, given in Table 6.6, are recommended for frequent monitoring in site C, and also in A.

Tailoring a safe and economically feasible monitoring program would also include the parameters which currently have low relevance, but could omit the question of metabolites at the moment. According to the risk matrix, the compounds diclofenac, ibuprofen, paracetamol, metoprolol, sotalol, clarithromycin, erythromycin, trimethoprim, bezafibrate, cetirizine, primidone, citalopram, the pesticides atrazine and diuron and the pesticide metabolites, desethylatrazine and desisopropylatrazine exhibit a low relevance at site A or C. Nevertheless, they should also be included in the monitoring program for Lake Paranoá but can sampled less frequently; a sampling at least once in every season is advisable.

Due to the insufficient elimination of many of the polar compounds during conventional drinking water treatment (Stackelberg *et al.* 2007), the analysis of polar and refractory emerging pollutants particularly in raw and drinking water is also highly recommended.

In case of a limited analytical capacity, it may be necessary to further reduce the number of compounds. Based on the results, the following substances are suggested as the most important anthropogenic key compounds for the water quality of surface water and drinking water in DF. The parameters are also suited for the selection of the proper technical options for drinking water treatment and in general to control the removal efficiency in upgrading existing WWTPs. These compounds can be found frequently in the lake either because of high amounts of usage in the DF and/or low removal efficiency in the existing WWTPs. The crucial compounds are caffeine, atenolol, phenazone, and iopromide. Additionally, sulfamethoxazole, bezafibrate, and gemfibrozil should be analysed. Instead of caffeine, the sweeteners aspartame, sucralose, or acesulfame can also be used as more stable anthropogenic tracers.

6.4.3.3 Stimulants and pharmaceuticals

Caffeine and its metabolites paraxanthine, theobromine, theophylline, 1-methylxanthine, and 3-methylxanthine and the nicotine metabolite cotinine were used to serve as general indicators for anthropogenic pollution. Besides caffeine and cotinine, only paraxanthine was found at all sampling sites. The other metabolites occur only sporadically with no clear trend. The concentration of caffeine itself seems to follow no clear seasonal or spatial tendency. Opposite to all other compounds investigated, concentrations decrease in the dry season and increase during the rainy season. Furthermore, the concentrations at all sampling sites are strongly fluctuating. Concentrations at site A are equal to the concentrations measured at the sites C and F, and are sometimes even lower. The fluctuating concentration of caffeine is possibly attributed to diffuse discharge into the tributaries. Misconnections to rainwater gullies could also be a reason. An increase during the rainy season could be due to rainwater runoff which reaches the lake, carrying loads from rural areas not connected to the urban sewage system while this is not the case during dry season.

From the investigated pharmaceuticals, only phenazone, atenolol, and sulfamethoxazole showed maximum concentrations above 100 ng L^{-1} in the lake. The highest concentrations of these compounds were reached in point A. A typical view on the distribution of the frequently occurring pharmaceuticals can be seen in Figure 6.9.

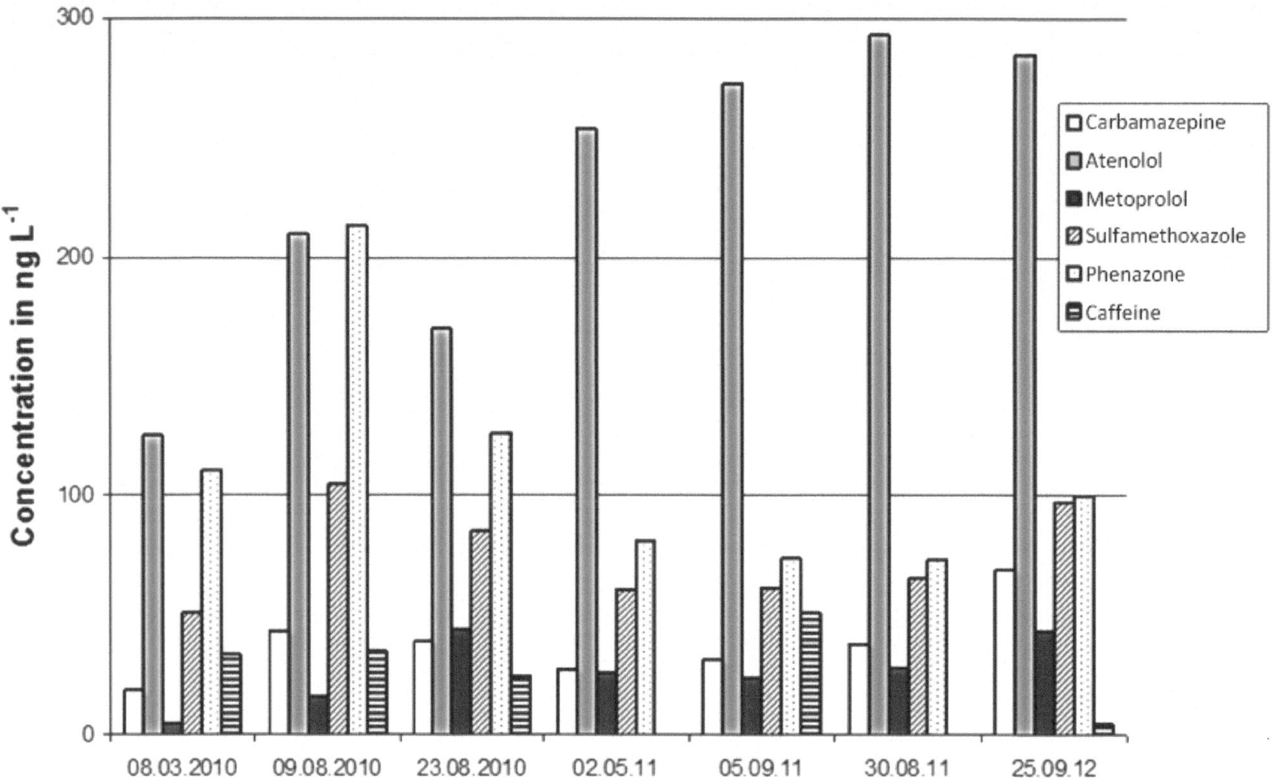

Figure 6.9 Mean concentrations of frequently occuring pharmceuticals and caffeine in Lake Paranoá (sampling site A), time period March 2010 to September 2012.

6.4.3.4 Pesticides

The pesticides propham, metamitron, 2-methyl-4-chlorophenoxyacetic acid (MCPA), acephat, primicarb, bentazon, and pyroquilon were analysed by LC-MS/MS in lake water and WWTP effluents. The concentrations of these substances were below the detection limit of the methods (LOQ) (<5.0 ng L^{-1} or <50 ng L^{-1} for WWTP effluent) in all samples. Since atrazine and its metabolite desethylatrazine was only detected sporadically in 2009 and 2010, it could be found in 100% of the lake samples in 2011 and 2012 between 2 to 14 ng L^{-1} and 2 to 22 ng L^{-1}, respectively. It is interesting to note that atrazine only occurred twice in the effluents of the WWTPs in concentrations of 15 and 8 ng L^{-1}. The herbicide diuron was found repeatedly at sampling site A ranging from 4 to17 ng L^{-1}.

A main input pathway for diuron appears to be the WWTP effluent with concentrations peaking up to 322 ng L^{-1}, most likely originating from public lawns, housing facades or park areas and being transported via surface water run-off into the sewer system. Furthermore, the discharge of Riacho Fundo could be a source for diuron, where the highest concentration was analyzed to be 79 ng L^{-1}. The metabolite desisoethylatrazine could only be detected at point A (2 to 14 ng L^{-1}). Despite the small-scale agricultural areas and the large amount of small parks and lawns across the capital, the measured herbicide concentrations in the lake are mostly close to the limit of detection (LOQ) of the methods, and they are clearly below the legally defined water classification thresholds of Brazil (CONAMA, 2005) as well as below the WHO guideline values (WHO, 2011). One reason for the low number of positive herbicide detections and their low concentrations could be the high soil retention capacity (Inoue *et al.* 2006). The detected herbicides and their metabolites seem to follow a seasonal trend, with a higher frequency of detection at the end of the rainy season.

6.4.3.5 Perfluorinated surfactants

Perfluorooctanoic acid (PFOA) and perfluorooctanesulfonate (PFOS) were measured as representatives for the group of perfluorinated compounds, which were added to the list of the Stockholm Convention on Persistent Organic Compounds in 2009. These substances are used in various applications such as textiles, fire-fighting materials, hydraulic fluids, carpet fabrics, photo industry or medical devices. Point sources can include fluorochemical facilities, carpet production or illegally recycled waste (Skutlarek *et al.* 2006).

Concentrations measured in lake water were found above the LOQ but lower than 2 ng L^{-1} ($n = 66$). PFOS is below the LOQ in all effluent samples ($n = 19$) except once in WWTP South with 3 ng L^{-1}. PFOA concentrations ranged between 4 and 74 ng L^{-1} in the effluents and was present above the LOQ in 50% of all samples taken. The lake water concentrations are considerably below the German drinking water guideline value of 300 ng L^{-1} (as sum of PFOS and PFOA; Dieter, 2011). Predicted no effect concentrations (PNEC) were reported with 50 ng L^{-1} (Rostkowski *et al.* 2006) and were generally not exceeded.

6.4.3.6 Artificial sweeteners

The five artificial sweeteners acesulfame, saccharin, aspartame, cyclamate and sucralose were included in the monitoring program. These substances are used as sugar substitutes in food, beverages and sanitary products. Their use in mass consumption products as well as their high water solubility and relatively high product concentrations make them particularly suitable as markers for anthropogenic wastewater in order to detect proportions of wastewater of >0.05% (Buerge *et al.* 2009).

Aspartame and cyclamate were not found in any of the lake samples between 2010 and 2012 ($n = 27$), while acesulfame and sucralose could be detected in all samples ranging from 9 to 265 ng L^{-1} (average: 75 ng L^{-1}) and from 49 to 255 ng L^{-1} (average: 144 ng L^{-1}), respectively. Saccharin occurred sporadically between 5 and 60 ng L^{-1}. However, no significant temporal or spatial variability could be observed for any of the compounds. The concentrations are in the same range as values reported for European surface waters with 100 to 920 ng L^{-1} for sucralose in 23 countries (Loos *et al.* 2009) or 400 to 1100 ng L^{-1} for acesulfame in the river Rhine (AWBR, 2011). While their ecotoxicological impact still needs to be investigated, the artificial sweeteners selected are not relevant for human toxicological assessment (Kroger *et al.* 2006).With regard to the WWTP effluents ($n = 15$), acesulfame concentrations ranged from 2300 to 18,100 ng L^{-1}, saccharin from 100 to 15,300 ng L^{-1} and sucralose from 100 to 4500 ng L^{-1} (Table 6.7). In contrast to the lake water, cyclamate could be detected in wastewater ranging between 100 and 1800 ng L^{-1}. The removal efficiency of conventional WWTPs for saccharin and cyclamate is usually >90%, while sucralose and acesulfame are not significantly removed and therefore, predestined as wastewater marker (Lange *et al.* 2012). With regard to Lake Paranoá, an increasing fraction of wastewater can be expected in the lake due to the rapid growth of the city and hence, continuous monitoring of these two indicators is advisable. Assuming no removal during wastewater treatment, for acesulfame and sucralose, the per capita consumption in mg per person and day can be calculated resulting in 3.1 for acesulfame and 0.9 for sucralose, which corresponds to values reported in literature for Europe (Lange *et al.* 2012).

Table 6.7. Measured concentrations of artificial sweeteners shown as minimum/median/maximum concentrations in µg L^{-1}; $n = 27$ in the lake; $n = 8$ for WWTP North; $n = 9$ for WWTP South.

	Acesulfam	Saccharin	Aspartame	Sucralose	Cyclamate
Lake A-F (ng L^{-1})	9 / 65 / 265	5 / 10 / 67	<LOQ	49 / 142 / 255	<LOQ
WWTP North (µg L^{-1})	6.1 / 14.0 / 18.1	0.1 / 4.1 / 14.9	<LOQ	2.3 / 3.1 / 4.5	0.1 / 0.3 /0.7
WWTP South (µg L^{-1})	4.1 / 14.8 / 17.9	0.6 / 5.5 / 15.3	<LOQ	2.1 / 2.9 / 3.5	0.1 / 1.1 / 1.8

6.4.3.7 X-ray contrast media

The X-ray contrast media diatrizoic acid, ioxithalamic acid, iopamidol, iomeprol, iopromide and iohexol were measured. These substances exhibit high stability towards degradation processes by the human metabolism and are excreted almost completely unchanged (Ternes & Hirsch, 2000).

Between 2010 and 2013, diatrizoic acid could be detected only in 15% of all lake samples ($n = 33$) with concentration from 20 to 25 ng L^{-1}. Ioxithalamic acid was present in 42% of all samples with concentrations between 50 and 590 ng L^{-1}. Iopamidol was only detected twice at 42 and 58 ng L^{-1} while iomeprol did not occur. Iopromide and iohexol could be found in 65% and 81% of the samples ($n = 70$) in a concentration range of 20 to 1320 ng L^{-1} and 20 to 960 ng L^{-1}, respectively. The temporal and spatial variation of these X-ray contrast media in the lake was very low except for iopromide (Figure 6.10). The health orientation value defined by the German Federal Environmental Agency (UBA) of 1000 ng L^{-1} per single x-ray contrast substance (IKSR, 2010) was exceeded only once (iopromide, $c = 1100$ ng L^{-1}).

The main input occurs via the WWTPs effluents, where concentrations varied to a large extent, for instance for ioxithalamic acid ranging from 0.1 to 22.2 µg L^{-1} (average = 8 µg L^{-1}). Iopromide and iohexol range from 0.29 to 14.5 µg L^{-1} with average concentrations ($n = 44$) of 3.8 µg L^{-1} and 1.9 µg L^{-1}, respectively. Iopamidol was detected with 2.8 µg L^{-1} on average ($n = 13$). High fluctuations can clearly be attributed to the specific punctual application of these compounds in X-ray imaging.

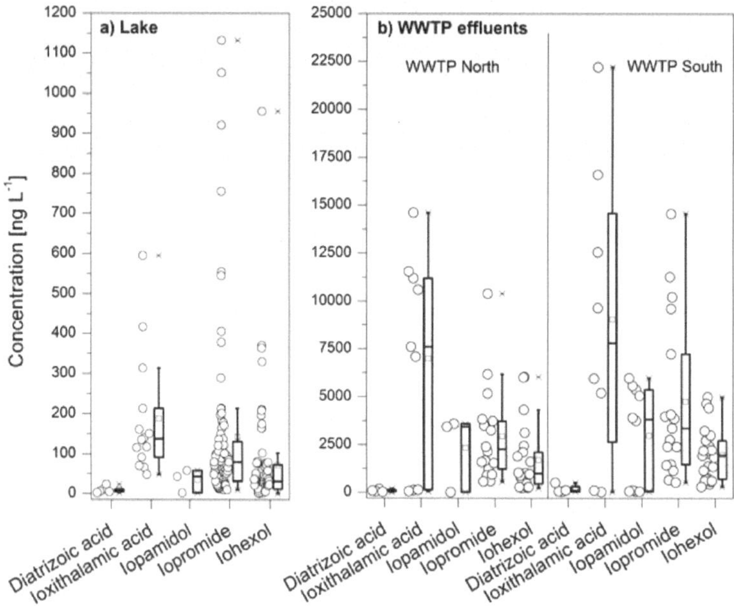

Figure 6.10 Measured concentrations of X-ray contrast media in Lake Paranoá and the WWTP effluents of WWTP North and South; upper and lower box margins = 25% and 75% percentiles; whisker = 1% and 99% percentiles; casket = average; horizontal line = median.

6.4.3.8 Other compounds

The hydrophilic corrosion inhibitors $1H$-benzotriazole and tolyltriazole are widely used in industry and as additives in household detergents. They show a relatively high stability in WWTPs and are therefore often detectable in European surface waters and groundwater up to 3.5 µg L^{-1} (Giger *et al.* 2006; Seeland *et al.* 2012). In the Lake Paranoá, both compounds were mostly determined in the lower ng L^{-1} range at all sampling sites; the highest concentration found was 34 ng L^{-1} for $1H$-benzotriazole and 140 ng L^{-1} for tolyltriazole, respectively, each at point A. In the time period from 2009 to 2012, no significant tendency for the triazole concentration could be observed. The concentrations of both corrosion inhibitors in WWTP effluents were much higher in comparison to the lake. The maximum concentrations were 221 ng L^{-1} for $1H$-benzotriazole and 814 ng L^{-1} for tolyltriazole. The content of tolyltriazole is always higher compared to $1H$-benzotriazole (see Figure 6.11).

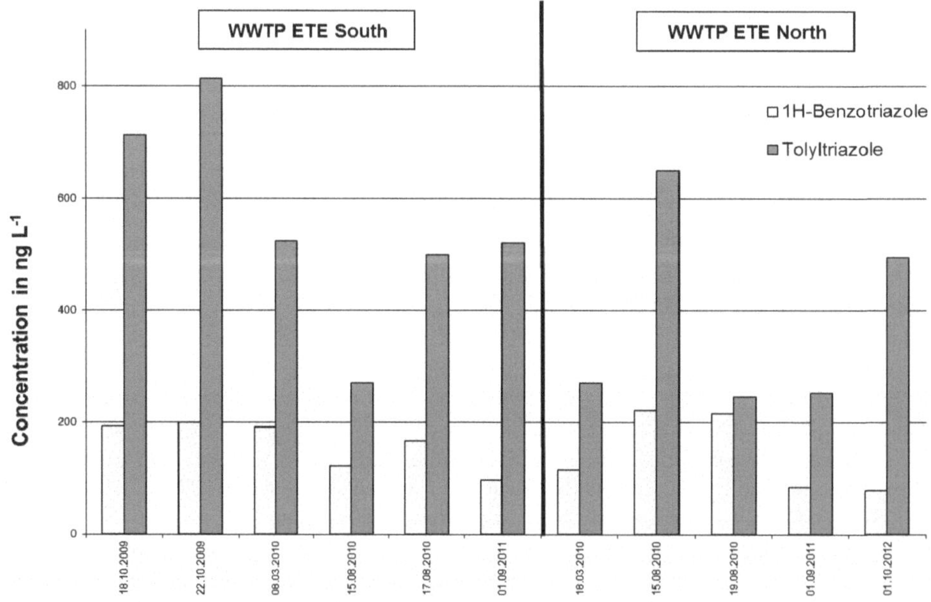

Figure 6.11 Concentration of the corrosion inhibitors 1H-benzotriazole and tolyltriazole in effluents of the both WWTP South and North, time period October 2009 – October 2012, mean values (n = 2–4).

The concentrations of the measured EPA polycyclic aromatic hydrocarbons (16 PAHs according to USEPA recommendation) did not exceed the LOQ in any of the aqueous samples in the lake and WWTP effluents (<0.1 µg L^{-1} in lake water and <1.0 µg L^{-1} WWTPs; measured using GC-MS). During the dry season in October 2009 and August 2010, the odorous compounds geosmin and 2-methylisoborneol were analyzed but found to be below LOQ of 4 ng L^{-1} despite the presence of cyanobacteria in Lake Paranoá (Steiniger *et al.* 2011), which are known to be possible producers of odorous compounds.

Microcystines-LR, -RR, -YR, -LF, -LW, and –dmRR: The concentrations of all targeted compounds were below the limit of detection (LOQ) (between <2 and 5 ng L^{-1}, depending on the enrichment factor) and below the WHO guideline value of 1 µg L^{-1} (Microcystine-LR; WHO, 1998). No cylindrospermopsin concentration was observed above the detection limit of the method (between <2 and 5 ng L^{-1}, depending on the enrichment factor).

6.5 CONCLUSIONS

This chapter presents the results of a comprehensive water quality assessment with specific focus on organic micropollutants, which was carried out for the first time for Lake Paranoá. With consideration of long-term existing data of CAESB as well as the information about the urban structure of the catchment area, the obtained results allow confirmation of a good to very good chemical quality of the lake water with respect to drinking water production in the future. This also holds for the ecotoxicological point of view concerning the basic water chemical parameters and the analyzed micropollutants. However, it should be noted that long-term monitoring for emerging polar micropollutants is indispensable, particularly because this has not been done to date.

Due to the discharge of two large WWTPs into the lake, persistent compounds can be expected to accumulate as similarly observed for European water bodies. The growing population will increase the pollutant load received by the lake and the natural self-purification of the lake will probably not be sufficient to eliminate these recalcitrant substances. As a matter of fact, monitoring of the substances proposed in Section 6.4.3 is strongly advisable in order to diagnose long-term trends. The set-up of the risk matrix based on measured concentrations of organic micropollutants provides the basis for decisions in view of the refinement of future monitoring campaigns. Monitoring is also recommended for the existing drinking water treatment plants, since numerous polar micropollutants are not removed in conventional drinking water treatment. The detailed plans should be frequently updated, and furthermore can be easily complimented with further parameters to suit changing needs. It also provides the basis for choosing a proper treatment technology for drinking water supply.

Concerning pesticides, atrazine and its metabolites as well as diuron have been detected. While atrazine and its metabolites occurred on a very low level, most likely due to leaching from residues in soils or groundwater, diuron showed higher concentrations and fluctuations. High proportions of pesticides can also originate from public urban lawns, house facades etc. and therefore, WWTP effluents should also be monitored and quantified in view of this aspect. A number of micropollutants could not be detected, and the measured concentrations were usually lower than the ones in similar European or North American lakes. For most compounds the concentrations were below 100 ng L^{-1}. It is attractive to consider an additional 4th step in the wastewater treatment to reduce the input into Lake Paranoá. For this enhanced removal, activated carbon might be an option. This is currently under investigation in the pilot plant installed at the WWTP North.

The DOC concentrations in Lake Paranoá are fairly low. Nevertheless, regarding the forecasted climate change, extreme events, further urbanization and increasing population growth, a continuous analysis of the sum parameter DOC or the surrogate parameter UV at = 254 nm is considered to be important. Using this parameter in its advanced form (e.g., by means of SEC-UV/DOC analysis), gradual tendencies of a changing water quality can be observed early and with relatively low effort.

None of the heavy metals measured in lake water exceeded Brazilian and German drinking water thresholds. This emphasizes the very good raw water quality. Sediments were not contaminated except for Cd and Pb but show only a very low remobilization potential. Anion concentrations also were found at non-critical levels, and the concentrations of total phosphorus and chlorophyll-a in the last years indicate that Lake Paranoá has reached a steady oligotrophic state.

However, the population growth and the challenges of the climate change might cause concern regarding the upkeep of this state. Hence, future trends should be closely and carefully monitored to be able to timely counteract negative developments. With regard to future drinking water production, it is a must to protect the raw water quality - the more, the better. This can be done by minimizing the input of diffuse and direct pollution and by optimizing the wastewater treatment. The responsibility should not be shifted over to the final drinking water treatment since the contamination of lake water with (newly emerging) substances or their transformation products, which possibly eludes the processes of drinking water treatment, could develop into a threat to the population. Furthermore, the protection of the whole regional ecosystem would guarantee a sustainable development of the aquatic ecosystem.

In view of further research, additional analyses are recommended to quantify the input via the tributaries and urban runoff during the rainy season. For instance, using passive sampling devices can help to obtain reliable mass balances for the constituents of Lake Paranoá water. Together with a deeper understanding of the limnological features, advanced modeling approaches can contribute to identify and understand the underlying processes in the lake as well as avoid further possible sources of organic micropollutants. A refined model can also be used to simulate different climate change influences and population growth scenarios leading to more precise recommendations, and in this way successfully support an integrated water management system.

6.6 ACKNOWLEDGEMENTS

The authors would like to thank Elly Karle and Axel Heidt for their commitment during the sampling campaigns in Brasilía, as well as Nicole Hebben, Rafael Peschke, Reinhardt Sembritzki, Matthias Weber and Ulrich Reichert for their fine analytical work in the laboratory and their project participation. Furthermore, the authors are grateful to Karsten Nödler and Tobias Licha from the Geoscience Centre of the University Göttingen for their analytical support. The authors want to thank the Federal Ministry of Education and Research (BMBF) for the financial support of the IWAS-ÁGUA DF Project (02WM1070, 02WM1028).

6.7 REFERENCES

Alder A. C., Schaffner C., Majewsky M., Klasmeier J. and Fenner, K. (2010). Fate of β-blocker human pharmaceuticals in surface water: comparison of measured and simulated concentrations in the Glatt Valley Watershed, Switzerland. *Water Research*, **44**(3), 936–948.

Altafin I. G., Mattos S. P., Cavalcanti C. G. B. and Estuqui V. R. (1995). Paranoá Lake – limnology and recovery program. In: Limnology in Brasil, J. G. Tundisi, E. M. Bicudo and E. M. Tundisi (eds.), Brazilian Academy of Sciences, Brazilian Limnological Society, pp. 325–349.

Angelini R., Bini L. M. and Starling F. L. R. M. (2008). Efeitos de diferentes intervenções no processo de eutrofização do Lago Paranoá (Brasília – DF) [Effects of interventions on the eutrophication process of Paranoá lake reservoir of Brasilia, Brazil]. Oecologia Brasiliensis, **12**(3), 564–571.

AWBR (2011). Arbeitsgemeinschaft Wasserwerke Bodensee Rhein e.V. (2011). 42. Jahresbericht 2010 [annual report working group waterworks Lake Constance Rhine] Geschäftsstelle der Arbeitsgemeinschaft Wasserwerke Bodensee-Rhein (AWBR), Freiburg, Germany.

Assis R. S. S. (2006). Remoção de Micocystis aeruginosa e microcistinas por flotação por ar dissolvido – Estudo de bancada utilizando sulfato de alumínio e cloreto ferrico como coagulante [Microcystis aeruginosa and microcistins removal by dissolved air flotation – bench scale study using aluminium sulphate and ferric chloride as coagulants]. Dissertação de Mestrado em Tecnologia Ambiental e recursos Hídricos. Universidade de Brasília. Brasila, DF. Brasil.

Bjork S. (1979). The Lago Parnaoá restoration project. Final report. Brasilia, Brazil. Project UNDP BRA/75/03 WHO BRA-2341.

Buerge I. J., Buser H.-R., Kahle M., Müller M. D. and Poiger T. (2009). Ubiquitous occurrence of the artificial sweetener acesulfame in the aquatic environment: an ideal chemical marker of domestic wastewater in groundwater. *Environmental Science & Technology*, **43**(12), 4381–4385.

Branco C. W. C. and Senna P. A. C. (1991). The taxonomic elucidation of the Paranoa Lake (Brasilia, Brazil) problem: Cylindrospermopsis raciborskii. Bulletin du Jardin botanique national de Belgique, **61**, 85–91.

Brasil C. P. (2004). Avaliação da remoção de microscistinas em água de abastecimento público por diferentes carvões ativados produzidos no Brasil [Evaluation of microscistins removal in public water supply by different activated carbons produced in Brazil]. Dissertação de Mestrado em Tecnologia Ambiental e recursos Hídricos. Universidade de Brasília. Brasila, DF. Brasil.

Bundschuh T., Knopp R., Winzenbacher R., Il Kim J. and Köster R. (2001). Quantification of aquatic nano particles after different steps of bodensee water purification with laser-induced breakdown detection (LIBD). *Acta hydrochimica et hydrobiologica*, **29**(1), 7–15.

Caldas E., Coelho R., Souza L. and Silva, S. (1999). Organochlorine pesticides in water, sediment, and fish of Paranoá Lake of Brasilia, *Brazil. Bulletin of Environmental Contamination and Toxicology*, **62**, 199–206.

Carballa M., Omil F., Lema J. M., Llompart M., García-Jares C., Rodríguez I., Gómez M. and Ternes T. A. (2004). Behavior of pharmaceuticals, cosmetics and hormones in a sewage treatment plant.*Water Research*, **38**(12), 2918–2926.

Carlson R. E. (1977). A trophic state index for lakes. *Limnology & Oceanography*, **22**, 361–369.

Cavalcanti C. G. B. (1988). Caracterizacão quimica preliminar do sedimento no Lago Paranoá, Brasília, Distrito Federal [Preliminary chemical characterization of the sediment from Paranoá Lake, Brasília, Distrito Federal]. *Proc. of the 2nd Congress of Brazilian Limnologists*, Cuiabá, MT.

Cavalcanti C. G. B., Pinto M. T., de Freitas H. J. and Moreira R. C. A. (1997). Paranoa Lake restoration: Impact of tertiary treatment of sewage in the watershed. *Verhandlungen - Internationale Vereinigung für theoretische und angewandte Limnologie*, **26**(2), 689–693.

CONAMA (2005). Resolução nº 357, de 17 de Março de 2005 National Council of Environment – Resolution 357, adopted on March 17, 2005.

Cronberg G. (1977). The Lago do Paranoá Restoration Project. Phytoplankton ecology and taxonomy: Final report. Brasília, Brazil. Project PAHO/WHO/77/WT/BRA/2341/04

Delay M. and Frimmel F.H. (2012). Nanoparticles in aquatic systems. *Analytical and Bioanalytical Chemistry*, **402**, 583–592.

Demoliner A., Caldas S. S., Costa F. P., Gonçalves F. F., Clementin R. M., Milani M. R. and Primel E. G. (2010). Development and validation of a method using SPE and LC-ESI-MS-MS for the determination of multiple classes of pesticides and metabolites in water samples. *Journal of the Brazilian Chemical Society*, **21**(8), 1424–1433.

Dianese J. C., Pigati P. and Kitayama K. (1976). Residues of chlorinated hydrocarbon pesticide in the lake of Brasilia.*Biologico*, **XLII**, 151–154.

Dieter H. (2011). Grenzwerte, Leitwerte, Orientierungswerte, Maßnahmenwerte - Aktuelle Definitionen und Höchstwerte [Thresholds, guiding values, orientation values, intervention values – Recent definitions and maximum thresholds], Umweltbundesamt. http://www.umweltdaten.de/wasser/themen/trinkwassertoxikologie/grenzwerte_leitwerte.pdf, acessed July 2013.

Doll T. E. and Frimmel F. H. (2004). Kinetic study of photocatalytic degradation of carbamazepine, clofibric acid, iomeprol and iopromide assisted by different TiO2 materials - determination of intermediates and reaction pathways. *Water Research*, **38**, 955–964.

Doll T. E. and Frimmel F. H. (2005). Photocatalytic degradation of carbamazepine, clofibric acid and iomeprol with P25 and Hombikat UV100 in presence of natural organic matter (NOM) and of other organic water constituents. *Water Research*, **39**, 403–411.

Drinking Water Ordinance of Germany (2011). from May 21st, 2001 (TrinkwV 2001), according to European Drinking Water Ordinance, 98/83/EC, November 3rd, 1998), updated May 3rd, 2011, implemented November 1st, 2011.

Eichhorn P., Rodrigues S. V., Baumann W. and Knepper T. P. (2002). Incomplete degradation of linear alkylbenzenesulfonate surfactants in Brazilian surface waters and pursuit of their polar metabolites in drinking waters. *Science of the Total Environment*, **284**(1–3), 123–134.

Elmoor-Loureiro L. M. A., Mendonca-Galvao L. and Padovesi-Fonseca C. (2004). New cladoceran records from Lake Paranoa, Central Brazil. *Brazilian Journal of Biology*, **64**(3A), 415–422.

European Commission (2000). Directive 2000/60/EC, Establishing a framework for community action in the field of water policy.European Commission PE-CONS 3639/1/100 Rev 1, Luxemborg.

Fent K., Weston A. A. and Caminada D. (2006). Ecotoxicology of human pharmaceuticals. *Aquatic Toxicology*, **76**, 122–159.

Förstner U. (2004). Sediment dynamics and pollutant mobility in rivers: An interdisciplinary approach. *Lakes & Reservoirs: Research and Management*, **9**, 25–40.

Frimmel F. H. and Abbt-Braun G. (2009). Dissolved organic matter (DOC) in natural environments. In: Biophysico-Chemical Processes Involving Natural Nonliving Organic Matter in Environmental Systems, P. M. Huang and N. Senesi (eds.), John Wiley & Sons, Hoboken, New Jersey, pp 367–406.

Frimmel F. H. and Abbt-Braun G. (2011). Sum parameters: potential and limitations. In: *Treatise on Water Science*, P. Wilderer (ed.), Oxford: Academic Press, vol. 1, pp. 3–29.

Frimmel F. H., Abbt-Braun G., Heumann K. G., Hock B., Lüdemann H.-D. and Spiteller, M. (eds.) (2002). *Refractory Organic Substances in the Environment*. Wiley-VCH, Weinheim.

Frimmel F. H. and Niessner R. (eds.) (2010). Nanoparticles in the Water Cycle. Springer-Verlag Berlin-Heidelberg.

Frimmel F. H. and Müller M. B. (eds.) (2006). Heil-Lasten. Springer Heidelberg.

Froehner S., Souza D. B., Machado K. S. and Rosa E. C. (2009). Tracking anthropogenic inputs in barigui river, Brazil using biomarkers. *Water, Air, and Soil Pollution*, **210**, 33–41.

Ghiselli G. (2006). Avaliação da qualidade das águas destina-das ao abastecimento publico na região de Campinas: ocorrência e determinação dos interferentes endócrinos (IE) e produtos farmacêuticos e de higiene pessoal (PFHP). Doctoral Thesis.University of Campinas. http://biq.iqm.unicamp.br/arquivos/teses/vtls000398476.pdf, accessed March 10, 2008.

Giger W., Schaffner C. and Kohler H.-P. E. (2006). Benzotriazole and tolyltriazole as aquatic contaminants. 1. Input and occurrence in rivers and lakes. *Environmental Science & Technology*, **40**(23), 7186–7192.

Gioia S. M. C. L., Pimentel M. M., Tessler M., Dantas E. L., Campos J. E. and Guimares E. M. (2006). Sources of anthropogenic lead in sediments from an artificial lake in Brasilia-central Brazil. *Science of the Total Environment*, **356**(1–3), 125–142.

Grisolia C. K. and Starling F. L. (2001). Micronuclei monitoring of fishes from Lake Paranoá under influence of sewage treatment plant discharges. *Mutation Research*, **491**(1–2), 39–44.

Huber S. A. and Frimmel F. H. (1991). Flow injection analysis of organic and inorganic carbon in the low-ppb range. *Analytical Chemistry*, **63**(19), 2122–2130.

Huber S.A. and Frimmel F. H. (1992). A liquid chromatographic system with multi-detection for the direct analysis of hydrophilic organic compounds in natural waters. *Fresenius Journal of Analytical Chemistry*, **342**, 198–200.

IBAMA, Brazilian Institute for the Environment and Renewable Natural Resources (2010). Pesticides and related products commercialized in Brazil in 2009: an environmental approach. Report, Brasília: Ibama, 2010.

IKSR Internationale Kommission zum Schutz des Rheins (2010). Auswertungsbericht Röntgenkontrastmittel [Evaluation report X-raycontrast media], http://www.iksr.org/uploads/media/IKSR-Bericht_Nr._187d.pdf,_accessed June 21, 2013.

Inoue M. H., Rubem S. Oliveira R. S., Jussara B., Regitano J. B., Tormena C. A., Constantin J. and Tornisielo V. L. (2006). Sorption desorption of atrazine and diuron in soilsfrom Southern Brazil. *Journal of Environmental Science and Health, Part B*, **41**, 605–621.

Jardim W. F., Montagner C. C., Pescara I. C., Umbuzeiro G. A., Di DeaBergamasco A. M., Eldridge M. L. and Sodré F. F. (2012). An integrated approach to evaluate emerging contaminants in drinking water. *Separation and Purification Technology*, **84**, 3–8.

Jeppesen E., Søndergaard M., Lauridsen T. L., Davidson T. A., Liu Z. Mazzeo N. Trochine C. Özkan K., Jensen H. S., Trolle D., Starling F., Lazzaro X., Johansson L. S., Bjerring R., Liboriussen L., Larsen S. E., Landkildehus F., Egemose S. and Meerhoff M. (2012). Biomanipulation as a restoration tool to combat eutrophication. *Advances in Ecological Research*, **47**, 411–488.

Joss A., Siegrist H. and Ternes T. A. (2008). Are we about to upgrade wastewater treatment for removing organic micropollutants? *Water Science and Technology*, **57**(2), 251–255.

Kuster M., Azevedo D. A., López de Alda M. J., Aquino Neto F. R. and Barceló D. (2009). Analysis of phytoestrogens, progestogens and estrogens in environmental waters from Rio de Janeiro (Brazil). *Environment International,* **35**(7), 997–1003.

Kroger M., Meister K. and Kava R. (2006). Low-calorie sweeteners and other sugar substitutes: A review of the safety issues. *Comprehensive Reviews in Food Science and Food Safety*, **5**(2), 35–47.

Kutschera K., Börnick H. and Worch E. (2009). Photoinitiated oxidation of geosmin and 2-methylisoborneol by irradiation with 254 nm and 185 nm UV light. *Water Research*, **43**(8), 2224–2232.

Kümmerer K. (2010). Pharmaceuticals in the environment. *Annual Review of Environment and Resources*, **35**(1), 57–75.

Lange F. T., Scheurer M. and Brauch H. J. (2012). Artificial sweeteners--a recently recognized class of emerging environmental contaminants: a review. *Analytical and Bioanalytical Chem*istry, **403**(9), 2503–2518.

Locatelli M. A. F, Fernando F. Sodre F. F., Wilson F. and Jardim W. F. (2010). Determination of antibiotics in Brazilian surface waters using liquid chromatography–electrospray tandem mass spectrometry. *Archives of Environmental Contamination and Toxicology*, **60**, 385–393.

Loos R., Gawlik B. M., Locoro G., Rimaviciute E., Contini S. and Bidoglio G. (2009). EU-wide survey of polar organic persistent pollutants in European river waters. *Environmental Pollution*, **157**(2), 561–568.

Maia P. D., Guimarães E. M., Moreira R. C. A. and Boaventura G. R. (2005). Estudo mineralógico dos sedimentos de fundo do Lago Paranoá, DF [Mineralogical study of bottom sediments of the Paranoá Lake (DF-Brazil)]. *Revista Brasileira de Geociências*, **35**(4), 535–541.

Maia P. D., Boaventura G. R. and Pires A. C. B. (2006). Distribuição espacial de elementos-traço em sedimentos do Lago Paranoá, DF [Spatial distribution of trace elements in Paranoá Lake sediments, DF]. *Geochimica brasiliensis*, **20**(2), 158–174.

Majewsky M., Farlin J., Bayerle M. and Galle T. (2013). A case-study on the accuracy of mass balances for xenobiotics in full-scale wastewater treatment plants. *Environmental Science: Processes & Impacts*, **15**(4), 730–738.

Maruoka M. T. S. and Nascimento E. L. C. (2006). Sources of anthropogenic lead in sediments from an artificial lake in Brasilia-central Brazil. *Science of the Total Environment*, **356**(1–3), 125–142.

Mattos S. P., Altafin I. G., de Freitas H. J., Cavalcanti C. G. B. and Alves V. R. E. (1992). Lake Paranoa, Brasilia, Brazil: Integrated management plan for its restoration. *Water Pollution Research Journal of Canada*, **27**(2), 271–286.

Ministério da Agricultura (1979). Relatório preliminar sobre poluição do Lago Paranoá. Superitendência do Desenvolvimento da Pesca, Programa de Pesquisa e Desenvolvimento Pesqueiro do Brasil, Brasilia. [Preliminary report on Paranoá Lake polution. Superintendence of Fisheries Development, Programme for Research and Development of Brazilian Fishing, Brasilia].

Müller G. (1986). Schadstoffe in Sedimenten - Sedimente als Schadstoffe [Contaminants in Sediments – Sediments as Contaminants]. Umweltgeologie-Band **79**, Wien.

Nödler K., Licha T., Bester K. and Sauter M. (2010). Development of a multi-residue analytical method, based on liquid chromatography–tandem mass spectrometry, for the simultaneous determination of 46 micro-contaminants in aqueous samples. *Journal of Chromatography A*, **1217**(42), 6511–6521.

Padovesi-Fonseca C. and Philomeno M. G. (2004). Effects of algicide (copper sulfate) application on short-term fluctuations of of phytoplankton in Lake Paranoá, Central Brazil. *Brazilian Journal of Biology*, **64**(4), 819–826.

Padovesi-Fonseca C., Philomeno M. G. and Andreoni-Batista C. (2009). Limnological features after a flushing event in Paranoá Reservoir, Central Brazil. *Acta Limnologica Brasiliensia*, **21**(3), 277–285.

Padovesi-Fonseca C., Mendonca-Galvao L. and Andreoni-Batista C. (2011). Rotifera, Paranoá reservoir, Brasília, Central Brazil. *Checklist*, **7**(3), 248–252.

Pires, A. C. B., Ianniruberto, M. (2008). Morphologic and stratigraphic characterization of Lake Paranoá.Scientific project, agreement FAPDF nº 33/2008 – Environmental technologies in collaboration with the Brasília office of rivers (Ministry of Shipping).

Raimundo C. C. M. (2007). Ocorrência de interferentes endócrinos e produtos farmacêuticos nas águas superficiais da bacia do rio Atibaia. [Occurrence of endocrines and pharmaceutical products in surface waters of catchment area of the river Atibaia] *Thesis*. Campinas State University.

Reemtsma T. and Jekel M. (eds.) (2006). Organic Pollutants in the Water Cycle. WILEY-VCH, Weinheim.

Reemtsma T., Weiss S., Mueller J., Petrovic M., González S., Barcelo D., Ventura F. and Knepper T. P. (2006). Polar pollutants entry into the water cycle by municipal wastewater: A European perspective. *Environmental Science & Technology*, **40**(17), 5451–5458.

Richardson, S. D. (2009). Water analysis: emerging contaminants and current issues. *Analytical Chemistry*, **81**(12), 4645–4677.

Rostkowski P., Yamashita N., So I. M. K., Taniyasu S., Lam P. K. S., Falandysz J., Lee K. T., Kim S. K., Khim J. S., Im S. H., Newsted J. L., Jones P. D., Kannan K. and Giesy J. P. (2006) Perfluorinated compounds in streams of the Shihwa industrial zone and Lake Shihwa, South Korea. *Environmental Toxicology and Chemistry,* **25**(9), 2374–2380.

Seeland A., Oetken M., Kiss A., Fries E. and Oehlmann J. (2012). Acute and chronic toxicity of benzotriazoles to aquatic organisms. *Environmental Science and Pollution Research*, **19**, 1781–1790.

Schindelin A. J. and Frimmel F. H. (2000) Nitrate and natural organic matter in aqueous solutions irradiated by simulated sunlight: Influence on the degradation of the pesticides dichlorprop and terbutylazine. *ESPR – Evironmental Science Pollution Research*, **7**(4), 205–210.

Schwarzenbach R. P., Escher B. I., Fenner K., Hofstetter T. B., Johnson C. A., von Gunten U. and Wehrli B. (2006) The challenge of micropollutants in aquatic systems. *Science*, **313**(5790), 1072–1077.

Schwoerbel J. (1993). Einführung in die Limnologie [An Introduction to Limnology]. 7th edition, UTB Gustav Fischer, Stuttgart, Jena.

Skutlarek D., Exner M. and Färber H. (2006). Perfluorinated surfactants in surface and drinking waters. *Environmental Science and Pollution Research*, **13**(5), 299–307.

Sodré F. F. Montagner C. C., Locatelli M. A. F. and Jardim W. F. (2007). Ocorrência de interferentes endócrinos e produtos farmacêuticos em águas superficiais da região de campinas (SP, Brasil*). Journal of the Brazilian Society Ecotoxicol*gy, **2/2**, 187–196.

Sodré F. F., Pescara I. C., Montagner C. C. and Jardim W. F. (2010a). Assessing selected estrogens and xenoestrogens in Brazilian surface waters by liquid chromatography–tandem mass spectrometry. *Microchemical Journal*, **96**(1), 92–98.

Sodré F. F., Locatelli M. A. F. and Jardim W. F. (2010b). Occurrence of emerging contaminants in brazilian drinking waters: A sewage-to-tap issue. *Water, Air and Soil Pollution*, **206**(1–4), 57–67.

Somlyody L. and Altafin I. (1992). Management of water-resources and eutrophication in the DF of Brazil. *Water Science and Technology*, **26**(7–8), 1813–1822.

Stackelberg P. E., Gibs J., Furlong E. T., Meyer M. T., Zaugg S. D. and Lippincott R. L. (2007). Efficiency of conventional drinking-water-treatment processes in removal of pharmaceuticals and other organic compounds. *Science of the Total Environment*, **377**(2–3), 255–272.

Starling F., Lazzaro X., Cavalcanti C. and Moreira R. (2002). Contribution of omnivorous tilapia to eutrophication of a shallow tropical reservoir: evidence from a fish kill. *Freshwater Biology*, **47**(12), 2443–2452.

Steiniger B., Börnick H., Hebben N., Abbt-Braun G., Frimmel F. H., Nödler K., Licha T. Cavalcanti C. G. B., Cavalcanti C. P. and Worch E. (2011). The occurrence of emerging organic pollutants in a tropical reservoir in Brazil – contributions to the management of Lake Paranoá. *Proceedings of the International Conference on Integrated Water Resources Management*, Dresden 2011.

Suffet I. H. and MacCarthy P. (eds.) (1989). Aquatic Humic Substances. Influence on Fate and Treatment of Pollutants. American Chemical Society, Washington DC.

Sumpter J. P. and Johnson A. C. (2005). Lessons from endocrine disruption and their application to other issues concerning trace organics in the aquatic environment. *Environmental Science & Technology*, **39**(12), 4321–4332.

Stumpf M., Ternes T. A., Rodrigues S. V., Wilken R.-D. and Baumann W. (1999). Polar drug residues in sewage and natural waters in the state of Rio de Janeiro, Brazil. *Science of the Total Environment*, **225**, 135–141.

Tercero Espinoza L. A., ter Haseborg E., Weber M., Karle E., Peschke R. and Frimmel F. H. (2011). Effect of selected metal ions on the photocatalytic degradation of bog lake water natural organic matter. *Water Research*, **45**, 1039–1048.

Ternes T. A. and Hirsch R. (2000). Occurrence and behavior of x-ray contrast media in sewage facilities and the aquatic environment. *Environmental Science & Technology*, **34**(13), 2741–2748.

Ternes T. A. (2001). Analytical methods for the determination of pharmaceuticals in aqueous environmental samples. *Trends in Analytical Chemistry*, **20**(8), 419–434.

Tixier C., Singer H. P., Oellers S. and Müller S. R. (2003). Occurrence and fate of carbamazepine, clofibric acid, diclofenac, ibuprofen, ketoprofen, and naproxen in surface waters. *Environmental Science and Technology*, **37**(6), 1061–1068.

Thurman E. M. (1985). Organic Geochemistry of Natural Waters.Martinus Nijhoff / Dr. W. Junk Publishers, Dordrecht.

Vasyukova E., Uhl W., Braga F., Simões C., Baylão T. and Neder K. (2011). Drinking water production from surface water sources in the tropics: Brasília DF, Brazil. *Environmental Earth Science*, **65**, 1587–1599.

Volpato D. E., de Souza B. V., Dalla Rosa L. G., Melo L. H., Daudt, C. A. S and Deboni L. (2005). Use of antibiotics without medical prescription. *Brazilian Journal of Infectious Diseases*, **9**(3), 288–291.

Walter T. and Petrere J. M. (2007). The small-scale urban reservoir fisheries of Lago Paranoá, Brasilia, DF, Brazil. *Brazilian Journal of Biology*, **67**(1), 9–21.

Wetzel R. G. (2001). Limnology: Lake and River Ecosystems. Academic Press, San Diego.

WHO (1998). Cyanobacterial toxins: Microcystin-LR in drinking-water. Background document for development of WHO Guidelines for Drinkingwater Quality. http://www.who.int/water_sanitation_health/dwq/chemicals/cyanobactoxins.pdf, accessed July 15, 2013.

WHO (2011). Guidelines for drinking water quality. 4th edition, World Health Organization, Geneva, Switzerland.

Zwiener C. and Frimmel F. (2004) LC-MS analysis in the aquatic environment and in water treatment technology – a critic review. Analytical and Bioanalytical Chemistry, **378**(4), 862–874.

Chapter 7

Bridging the gap: Current and future drinking water treatment for a fast-growing megacity – Brasília, Distrito Federal

E. Vasyukova, K. R. F. O. Dassan, F. Braga, C. Simões, T. Baylão, K. Neder and W. Uhl

7.1 INTRODUCTION

The implementation of Integrated Water Resources Management (IWRM) is an essential element for managing water resources more sustainably, leading to long-term social, environmental and economic benefits. Two of the most important goals of IWRM involve bridging the deficit between water availability and demand, thus minimising the risk of water shortages and increasing the reliability of the water supply. In this respect, more attention should be paid to developing and saving water resources in order to meet the increasing demand, protecting degrading water resources and improving or upgrading existing treatment for potable water production.

The main natural feature challenging water production in the tropics is the large daily variation in rainfall volume and, consequently, the variable quality of surface water resources. This is seen in the fluctuations of particle concentrations in rivers and reservoirs. Such, sometimes sudden, changes in water quality may require very fast adaptation of the treatment process, which may need to be reconsidered with time (e.g., in terms of increasing the use of different chemical products or combinations thereof). Among the most important anthropogenic challenges are population growth and land use alterations, which typically influence surface water quantity and composition.

The Distrito Federal (DF) in central Brazil has been experiencing such pressures for several decades (Lorz *et al.* 2012; Strauch *et al.* 2013, Chapter 4, this book). Furthermore, considerable population growth (20% between 2000 and 2007 in the DF) has inevitably led to peripheral settlement expansion, which in turn has resulted in the longer residence time of drinking water in the distribution system. Due to this fact, the additional influence on tap water quality from various physical, chemical and bacteriological reactions must be taken into account in the future. In conjunction with the extreme differences in water availability between the dry and rainy seasons which are typical of the outer tropics, there is also the risk of a potential drinking water deficit in the DF in years to come. This possibility has already driven local authorities and water supply companies to increase water production from alternative water sources, although these are very limited.

The main goals of the IWRM concept developed for the DF of Brasília within the IWAS-ÁGUA DF project are to provide a tool for the assessment and management of all factors influencing the quality and quantity of available surface water resources, as well as to make technical and economic efforts aimed at ensuring a safe and sustainable water supply in the future (Aster *et al.* 2010; Kalbus *et al.* 2012). The objective of the present study was to contribute to the IWRM concept by reviewing the current state of drinking water treatment and supply in the capital city and suggesting possible ways of improving future water supply.

7.2 REVIEW OF EXISTING DRINKING WATER TREATMENT

7.2.1 Water supply system in the DF

Surface water is the most important freshwater resource for drinking water production in the DF. The two largest reservoirs, the Santa Maria and Descoberto (Figure 7.1), contain water of very good quality and satisfy almost 90% of water demand in the district. The remainder is provided by smaller reservoirs, streams and groundwater, which supply the systems of Planaltina-Sobradinho, Brazlândia and São Sebastião.

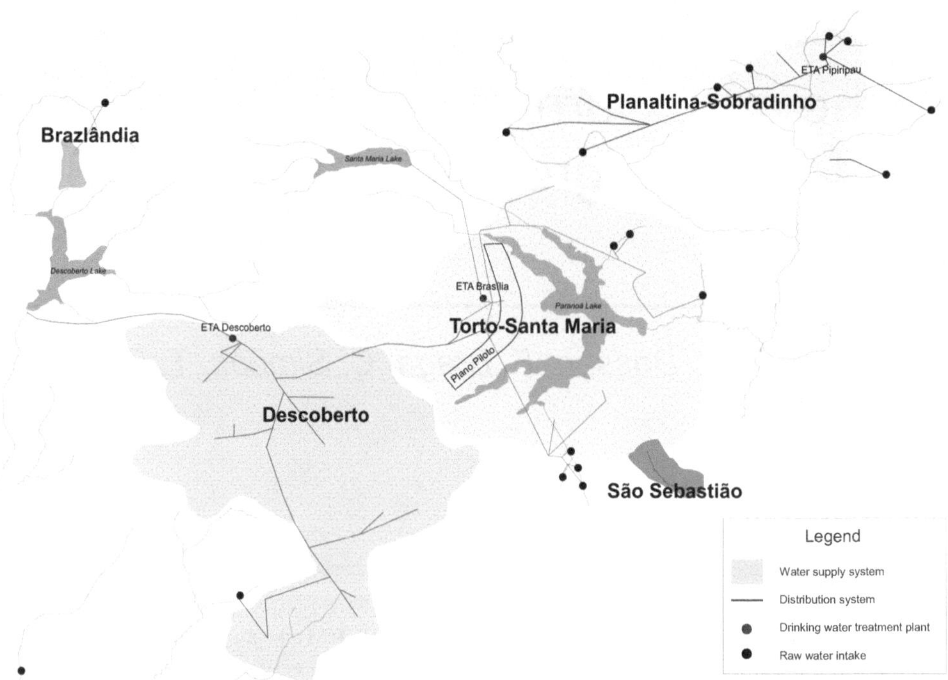

Figure 7.1 Water supply systems of the DF of Brasília (after the local water supply and sewerage company CAESB).

In 2008, the average production capacity of the entire water supply system reached 33,000 m^3/h, serving 99% of the urban population of the DF. The production capacities of the three major water supply systems are presented in Table 7.1.

Table 7.1 Main water supply systems in the DF.

Water supply system	Availability m^3/h	Mean flow in 2008 m^3/h	Supplied population %	Of total supply of CAESB %
Rio Descoberto	23810	15484	65	61
Torto/Santa Maria	8010	7189	17	27
Planaltina/Sobradinho	4303	2272	17	9

The present drinking water supply situation in Brasília was evaluated based on the example of three main waterworks located within the district, with each of these facilities obtaining their raw water from different types of surface water source. Waterworks A (WWA) treats reservoir water (named reservoir A in this work) via coagulation/flocculation with aluminium sulphate and direct sand filtration. Waterworks B (WWB) obtains water from another reservoir (B), but also from a river (B) situated in a protected zone. This mixture of river and reservoir water is treated via coagulation/flocculation using iron (III) chloride, dissolved air flotation and sand filtration. Surface water originating from three rivers (C-1, C-2 and C-3) is treated at waterworks C (WWC). The treatment process here includes coagulation/flocculation with aluminium sulphate, as well as upflow followed by downflow direct filtration (double-stage filtration).

7.2.2 Water quality and influence of weather conditions

Surface waters of the DF of Brasília are mainly neutral to slightly acidic (pH 6.3 to 7.1 on average) and are characterised by low mineralisation (conductivity <73 µS/cm), as well as highly variable turbidity (1 to 180 NTU), apparent colour (1 to >100 mg Pt/L) and microbiological parameters (i.e., total coliforms 2/100 mL to >2420/100 mL). At the same time, heavy metal and trace organic substance contents are low or below the detection limits. Elevated phytoplankton levels in the two main lakes (A and B) are caused by the presence of algae throughout the year. However, all monitored parameters in the water bodies are compliant with the state regulation (CONAMA, 2005).

The main peculiarity of the district's tropical climate is its uneven rainfall distribution, every year is comprised of both dry and rainy seasons, which typically last from May to October and from November to April, respectively. Based on the variation in water quality parameters, each year can be similarly divided into two periods of low and high particulate matter concentrations (Figure 7.2). Another specific feature characterising the tropical climate is the abundance of individual heavy rain events. Typically, 90% of daily rainfall in the DF is below 10 mm, although in rare instances levels can exceed 100 mm. In addition, the uneven distribution of rainfall across the DF is typified by the fact that one water source catchment may experience a high daily amount of rainfall (>100 mm) and the others not. Water turbidity during the rainy season can increase by up to 50–180 NTU and apparent colour by up to 70–100 mg Pt/L, with rivers the most subject to such fluctuations (Figure 7.2). It should be noted that the apparent colour data presented in this paper are for unfiltered samples and thus correlate well with turbidity (Vasyukova *et al.* 2012). Such increases in turbidity and apparent colour may be caused by rainfall-induced erosion, especially given that the lake tributaries and river C-1, for example, drain areas of intensive agriculture (Hurst *et al.* 2004; Lee *et al.* 2009). Other water sources in the DF exhibit moderate seasonal variations, with turbidity ranging from 1 to 36 NTU and apparent colour from 3 to 40 mg Pt/L. Generally it can be stated that fluctuations in these two parameters are more pronounced in rivers and less so in lakes due to the buffering capacity of the latter.

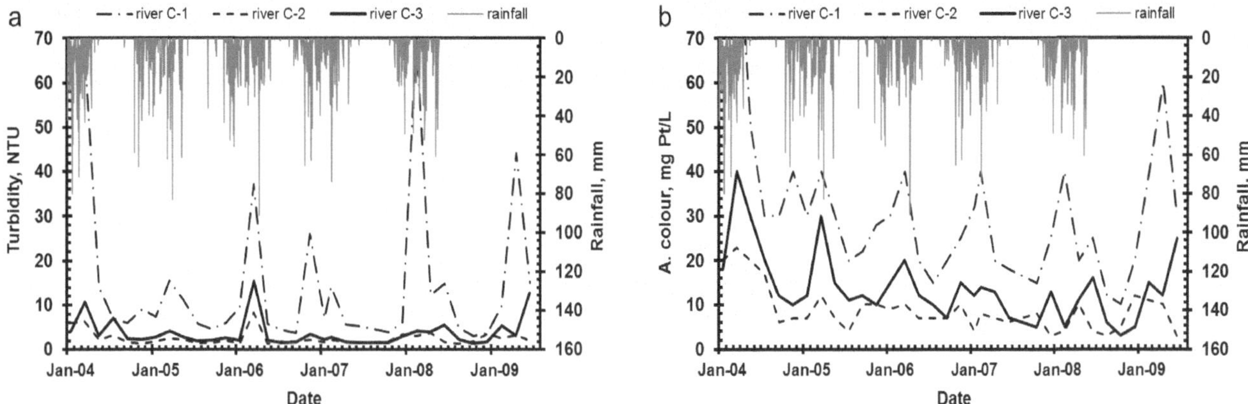

Figure 7.2 Monitored turbidity, apparent colour and rainfall values versus time for rivers C-1, C-2 and C-3, for the period 2004–2009. CONAMA (2005) regulation threshold levels for (a) raw water turbidity and (b) apparent colour correspond to 100 NTU and 75 mg Pt/L, respectively.

Extreme changes in surface water particulate matter concentrations thus represent the primary challenge facing waterworks in the DF. This is especially true during the rainy season, when (1) the quality of the water to be treated at one facility varies by one order of magnitude; and (2) raw water quality is significantly different from that observed during the dry season. Such problems in turn necessitate rapid adjustment of the treatment process, notably the coagulation procedure, and may lead to degradation of drinking water quality as a result.

7.2.3 Removal of particulate matter

Brazilian drinking water standards issued by the Brazilian Ministry of Health set threshold levels for drinking water turbidity and apparent colour at 5 NTU and 15 mg Pt/L, respectively (Portaria, 2004). The drinking water produced by the waterworks during the investigated period (January 2000 to August 2009) complied with these national regulations, although removal effectiveness was found to vary considerably.

Figure 7.3 presents the percentages of turbidity and apparent colour removal versus their initial values in raw water at WWA and WWC. Treatment effectiveness at WWA varied between 85 and 97% for the removal of both parameters, with values at their maximum and most stable when raw water turbidity and apparent colour were in the range of 5–18 NTU and 20–40 mg Pt/L, respectively. Above these levels, treatment effectiveness exhibited a slight tendency to decrease. In the case of WWC, the removal of both parameters was almost always maintained at around 90–98%, demonstrating the success of the employed treatment process – even for the mixed raw water originating from three different rivers. However, similar to the pattern observed at WWA, a slight decrease in apparent colour removal was observed when levels in raw water exceeded 40 mg Pt/L. It is interesting to note that in order to avoid pump damage potentially caused by a high suspended solids content in the water intake, the maximum turbidity level at WWC is currently set to 100 NTU for the river-sourced intake. A stricter limitation also exists for a smaller river in the same water supply system (not discussed here), where pumped raw water turbidity should not exceed 40 NTU. At almost all treatment plants a positive correlation between turbidity and apparent colour was observed in both raw ($r^2 > 0.6$) and treated waters ($r^2 > 0.7$), supporting the fact that the apparent colour of the source waters originates largely from particulate matter. Whereas WWA removed both parameters by a similar percentage, a poor correlation between the removal of the two parameters was observed at WWC ($r^2 = 0.4$). The fact that the raw water treated by WWA originates from only one source (reservoir) and is thus of a more predictable quality than the mixture of sources exploited by the other waterworks, may help WWA to tackle the challenge of rapid adjustment to changing lake water quality.

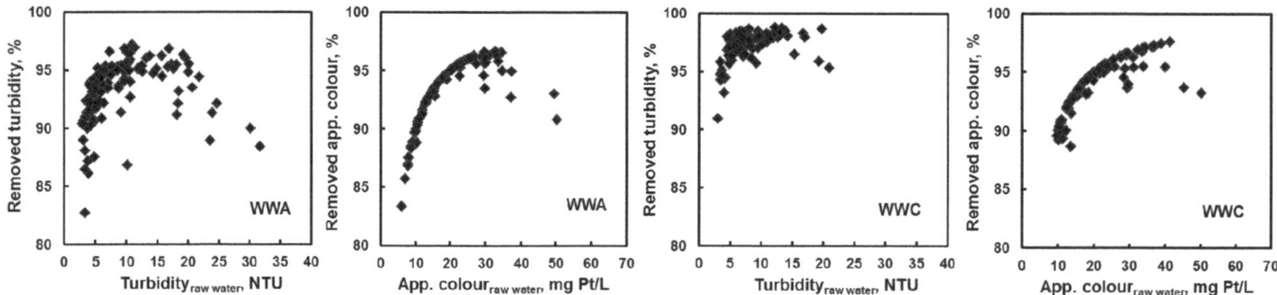

Figure 7.3 Percentage of turbidity and apparent colour removal versus turbidity and apparent colour in raw water at WWA and WWC, respectively.

In contrast to that for the two waterworks mentioned above, the plot of turbidity versus apparent colour in both raw and treated water samples at WWB reveals a poor correlation ($r^2 < 0.3$; data not shown), suggesting that water colour here is likely to (at least partially) originate from the washout of dissolved organic substances from upper soil horizons during rainfall events (Fellman *et al.* 2009; Morel *et al.* 2009). Moreover, the poor relationship in terms of their percentage removal ($r^2 = 0.4$) suggests that the mechanisms, and thus also the optimum conditions, for turbidity and apparent colour removal at the facility differ. Indeed, as two individual parameters, turbidity and natural organic matter removal is not always successful under the same conditions (Semmens & Field 1980; Gregor *et al.* 1997). The most desirable outcome is therefore to achieve a compromise between the respective optimum conditions for the removal of turbidity and organic substances via coagulation.

7.2.4 Removal of dissolved organic matter

As there is currently no legislative framework in place requiring the studied waterworks to continuously monitor organic substances and measure integrated parameters such as UV absorption at a wavelength of 254 nm (UVA_{254}) or dissolved organic carbon (DOC), DOC removal, including that of specific fractions, was evaluated at each of the three facilities using samples obtained in October 2009 and August 2010 (dry season), and April and November 2010 (rainy season). Bulk and fractional DOC were analysed via size exclusion chromatography with organic carbon detection (LC-OCD, Huber *et al.* 2011) before and after each treatment step at each of the considered waterworks. All raw waters were characterised by low dissolved organic carbon concentrations below 1.3 mgC/L. In the data available (Table 7.2), no clear seasonal pattern in raw water DOC levels can be observed. Waterworks C exhibited the largest variations, with DOC ranging from 0.54–1.12 mgC/L in raw water and 0.52–1.26 mgC/L in filter effluent. Such fluctuations may be explained by the fact that river C-1 drains an area characterised by intensive agriculture (Jardé *et al.* 2007).

Table 7.2. DOC concentrations (in mgC/L) of treatment trains at WWA, WWB and WWC for the period October 2009 to November 2010, n = number of samples.

Sample	WWA (n = 5)				WWB (n = 8)				WWC (n = 6)			
	average	min	max	std.dev.	average	min	max	std.dev.	average	min	max	std.dev.
Raw water	0.74	0.63	0.92	0.12	1.01	0.82	1.16	0.15	0.90	0.54	1.12	0.28
Coagulated water	0.64	0.55	0.75	0.07	0.55	0.48	0.60	0.05	0.84	0.41	1.43	0.40
Clarified water right	–	–	–	–	0.59	0.49	0.70	0.08	–	–	–	–
Clarified water left	–	–	–	–	0.54	0.44	0.61	0.06	–	–	–	–
Upflow sand filt. water	–	–	–	–	–	–	–	–	0.75	0.45	1.13	0.23
Downflow sand filt. water	0.63	0.49	1.05	0.19	0.54	0.43	0.61	0.08	0.78	0.52	1.26	0.26

Typical for surface waters, humic substances were the dominant fraction in the raw water samples from all three treatment plants, accounting for an average of 30 to 53% of DOC (Figure 7.4). The molecularity of these humic substances was rather low and varied from 600 to 700 g/mol on average, reflecting their natural position between aquagenic and pedogenic fulvic acid in type, with the latter more pronounced in the river waters treated at WWC (not shown). Biopolymers were detected in all raw water samples and represented 5 to 28% of total DOC, with the highest percentage recorded at WWB, likely due to the presence of algal organic matter in the reservoir (Henderson *et al.* 2008).

For all treatment plants, the contribution of building blocks and HOC to bulk DOC was similar, accounting for on average 16–22% and 7–11% of DOC, respectively. Levels of low molecular weight acids were below the detection limit. DOC was removed mainly via coagulation, with the process accounting for an average of 46% at WWB, but only 13% and 7% at WWA and WWC, respectively (for a detailed discussion see Vasyukova *et al.* 2012). At all facilities, high and medium molecular weight fractions, that is, biopolymers and humic substances, were preferentially removed, which is in accordance with previously published studies examining the treatment of surface waters (e.g., Amy *et al.* 1992). Building blocks and low molecular weight fractions exhibited no significant change during treatment within ±20% at all three facilities. The observed variation in DOC removal efficiencies between the waterworks might be caused by a number of factors, such as coagulant type and dose, as well as coagulation pH, the details of which were unavailable. However, the recorded overall low DOC removal effectiveness is more likely to be related to low bulk DOC concentrations in raw water.

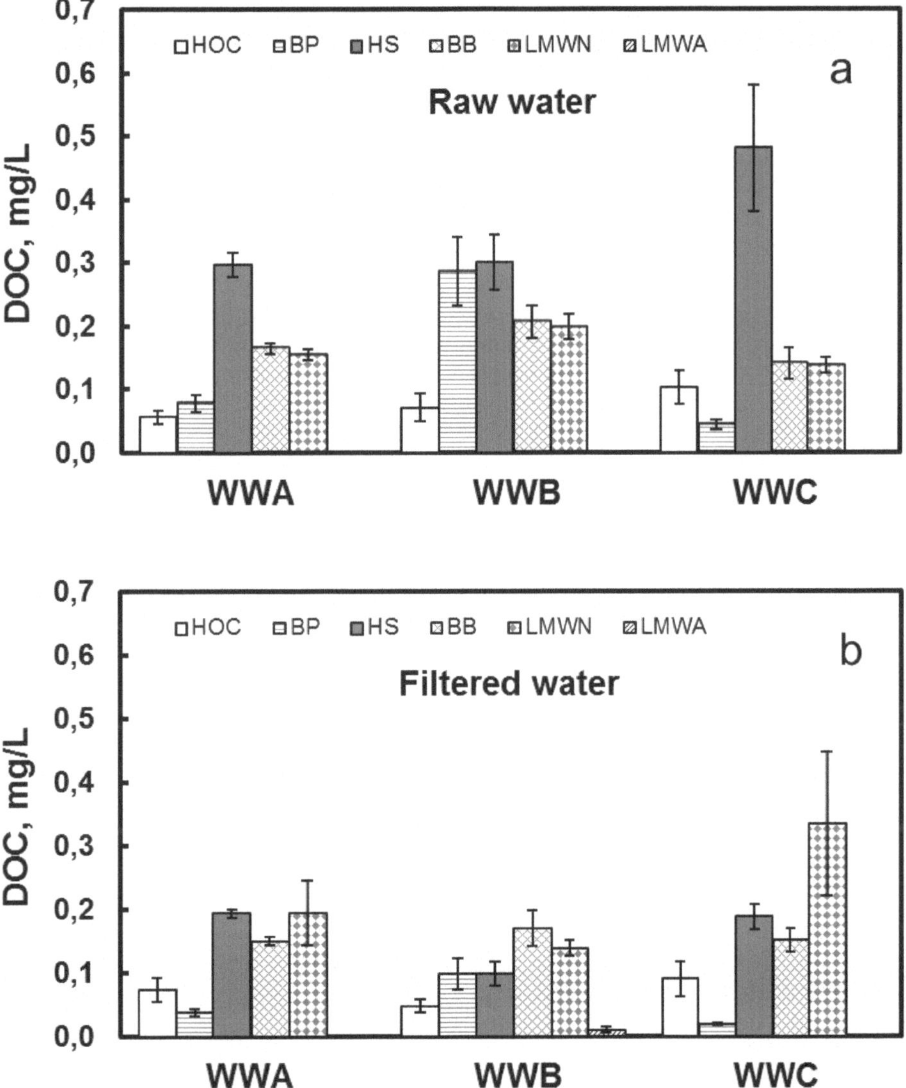

Figure 7.4 Mean fractional DOC composition of raw (a) and filtered (b) water at WWA, WWB and WWC. Error bars represent 67% confidence interval. HOC = hydrophobic organic carbon, BP = biopolymers, HS = humic substances, BB = building blocks, LMWN and LMWA = low molecular weight neutrals and acids.

Generally, it can be concluded that the surface waters of the DF are relatively well suited for treatment based on their natural organic matter content, with the treatment plants typically achieving low DOC concentrations in the treated water (Figure 7.4). However, the elevated biopolymer concentrations recorded at WWB (up to 0.35 mgC/L) may cause problems with the coagulation of particulate matter (Mania, 2006). Such substances are also assumed to contribute to biofilm formation and bacterial regrowth in distribution systems and the majority should therefore be removed (Vu *et al.* 2009). Furthermore, humic substances of the aquagenic fulvic acid type are generally more difficult to remove via coagulation than those of pedogenic fulvic acid origin (Huber *et al.* 2011). Thus the coagulation-based removal of humics at waterworks A and B was not as efficient as one would expect. Although the observed humic substance concentrations are low (<0.5 mgC/L), treatment should aim for the maximal removal of high and medium molecular weight organic compounds in order to minimise the potential for disinfection by-product formation due to chlorination (Singer, 1999).

7.3 TOWARDS A SUSTAINABLE WATER SUPPLY IN THE FUTURE

7.3.1 Possible upgrades for the better performance of conventional technologies

The results presented above demonstrate that seasonal and daily variations in rainfall play an important role in influencing the quality of the surface water sources used for drinking water production in the DF. In most surface sources, particles and apparent colour are likely to originate from erosion during rain events. This represents the primary challenge facing the waterworks, notably when the raw water qualities of the individual sources to be treated at each facility differ by one order of magnitude, as well as significantly varying from one season to the next. Drinking water quality was also found to be significantly influenced by raw water quality, that is, a decrease in removal efficiencies was observed during the wet season, with the turbidity of raw water >20 NTU and apparent colour >40 mg Pt/L at the three investigated waterworks. This suggests that (1) the coagulation procedure has the potential for optimisation, and (2) conventional treatment might not be enough to both provide a reliable barrier under variable raw water quality conditions and fulfil legal drinking water requirements. Furthermore, it was demonstrated that at all plants it is difficult to optimise the coagulation process for the simultaneous removal of particles and natural organic matter, including its specific fractions, due to rapid changes in raw water composition.

Several possibilities can be suggested which may improve the situation. The first guarantee of a more stable process involves the support of treatment via analytical and monitoring methods (Hurst *et al.* 2004). One such support measure is the introduction of online turbidity monitoring which enables rapid reaction to changing raw water quality, thus optimising the coagulation process for the removal of particulate matter. Such an online system has already been implemented at one of the existing waterworks, resulting in improved stability in terms of turbidity and apparent colour removal by up to 99%. Furthermore, the introduction of regular total or dissolved organic carbon and/or UVA_{254} measurements for organic matter monitoring could also be advantageous, especially if minimisation of potential disinfection by-product formation is pursued.

In addition to monitoring, coagulant dose prediction should be considered for the reliable removal of both turbidity and DOC. An alternative to laborious and time consuming jar-tests, techniques such as streaming current titration have proven suitable for rapid and accurate determination of the coagulation dosage, and are especially useful for highly turbid waters processed in water treatment plants, even during the rainy season (Dentel *et al.* 1989; Nam *et al.* 2013). Other studies have demonstrated that measurement of floc characteristics, for example, via the use of a photometric dispersion analyser, can be a useful tool with which to optimise the coagulation procedure (Briley & Knappe 2002; Staaks *et al.* 2011). Furthermore, the application of a more sophisticated control strategy, such as online control using artificial neural networks, may help to improve process safety and stability, as well as reduce treatment costs (Maier *et al.* 2004; Mälzer & Strugholz, 2008; Wu & Lo, 2008).

Finally, the introduction of membrane filtration to the conventional treatment train should be mentioned. Amongst the currently-available low-pressure membrane technologies, micro- and ultrafiltration could provide a risk-free alternative in cases of highly variable particle and microorganism content (Mierzwa *et al.* 2012). Although treatment costs would then be somewhat higher compared to multilayer filtration, the result would be a more constant drinking water quality with no fluctuation in turbidity, including in terms of microorganisms and apparent colour, a feat hard to achieve in tropical climates characterised by fast-changing weather conditions.

7.3.2 Expanding the water supply capacity

7.3.2.1 Lake Paranoá as a possible solution

The average metered water consumption in Brasília in 2003 was ca. 195 L per capita per day. However, factors including increased urbanisation and income levels, high water subsidies, increased landscaping and the construction of private swimming pools, have recently led to an increase in municipal water demand to a maximum average of 597 L per capita per day (data supplied by the water supply company). As a result the district's water supply and distribution system has been exposed to permanent pressure to adapt to these influences over the past few decades, with the existing waterworks already reaching their production capacity limits during periods of drought. By 2030 the population of the DF is expected to be around 3.5 million and thus the risk of future water shortages is imminent. Studies performed for the 'Plano Diretor de Água e Esgotos' (Water and Sewage Master Plan, CAESB, 2004) of the local water supply company have revealed the need to increase system production capacity from the current average of 6.1 m³/sec to 12 m³/sec by 2030, using new sources of water supply located in or even outside the DF (Altafin & Azevedo, 2007).

The DF is located in a region of springs, with a low level of surface water availability. The only source situated entirely within the District considered suitable for the extension of potable water supply on a short-term basis is currently the artificial Lake Paranoá (see Chapter 6, this book). A project examining water intake from Lake Paranoá has suggested the enlargement of existing system capacity by 2.8 m³/sec, which would provide approximatly additional 500,000 people with potable water. However, this option would pose certain challenges with respect to treatment. Embraced by the city of Brasília, Lake Paranoá has been subject to multiple pressures since its creation, including recreational activities, urban and agricultural runoff, as well as the discharge of treated secondary effluent from two waste water treatment plants (see Chapter 8, this book). According to a thorough investigation of lake water quality carried out as part of the present project, the main parameters of concern for potable water production are the presence of particles including algae and microorganisms, as well as emerging trace pollutants such as endocrine-disrupting compounds, pharmaceutically-active compounds, pesticides and personal care products (Chapter 6, this book; Hebben *et al.* 2011; Steiniger *et al.* 2011; Lorz *et al.* 2012).

7.3.2.2 Technological considerations for the treatment of Lake Paranoá water

At first, a treatment scheme based on conventional technologies was considered for lake water treatment. This scheme included pre-chlorination of raw water, coagulation/flocculation, flotation, multilayer filtration, and hybrid UV disinfection and chlorination (Figure 7.5). The possibility of adding ozonation followed by adsorption on the granular activated carbon after multilayer filtration during the second construction phase was also put forward. However, there are several arguments in favour of the employment of advanced treatment technologies.

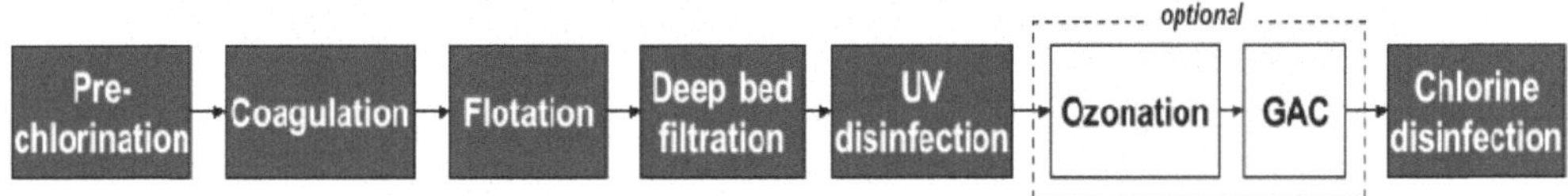

Figure 7.5 Original scheme for Lake Paranoá water treatment based on conventional technologies. UV stands for ultraviolet and GAC for granular activated carbon.

First, as is currently observed in existing waterworks, inconsistent particulate matter removal would be expected if only conventional treatment techniques are installed, due to the observed turbidity fluctuations in Lake Paranoá (unpublished). Substitution of the flotation and filtration steps by micro- or ultrafiltration may be an alternative with which to improve process stability under such conditions of variable raw water turbidity (Figure 7.6).

Second, if there is neither a target treatment installed at both waste water treatment plants on Lake Paranoá, nor complete collection of urban run-off aimed at avoiding the discharge of emerging trace contaminants into the water body, these contaminants should be removed as far as possible during drinking water treatment. It is known that the fate and transport of organic micropollutants (OMPs) depend on their properties (Siegrist & Joss, 2012). However, due to the wide range of the latter, such substances are difficult to analyse, with many remaining unidentified and their toxicity not yet known. In the past few decades, OMPs have been reported as having potentially adverse effects on human health (Bedding *et al.* 1983; Cotruvo, 1985). Furthermore, some OMPs can undergo transformation due to metabolism or oxidation/disinfection, with their by-products also posing a possible health risk.

With respect to OMP removability, treatment processes vary in effectiveness depending on the trace contaminant in question. Although clarification and sand filtration can considerably reduce particulate and dissolved matter levels, their effect is almost negligible regarding the removal of most OMPs detected in Lake Paranoá and waste water treatment plant effluent (Heberer *et al.* 1998; Ternes *et al.* 2002; Stackelberg *et al.* 2004, 2007; Loraine & Pettigrove, 2006; Rigobello *et al.* 2013). Only non-polar substances such as cholesterol can be removed via the coagulation of particles on which they adsorb, or via direct adsorption onto the membrane surface (POSEIDON, 2004; Stackelberg *et al.* 2007).

Bank filtration or artificial groundwater recharge using a ground passage generally results in the effective reduction of micropollutant concentrations via biodegradation or retention (Grünheid *et al.* 2005; Maeng *et al.* 2013). For this reason, bank filtration is preferred to the direct extraction of water from rivers in a number of countries. However, the abundant presence of private sector land-use along the lake shores limits or even excludes this possibility for the DF.

With recorded removal rates of >90%, biofiltration has proven to be effective in dealing with a number of pharmaceuticals and personal care products, including ibuprofen and paracetamol (Reungoat *et al.* 2011). However, certain common pharmaceuticals such as carbamazepine and diclofenac, as well as hormones such as estradiol and steroid hormones, are more persistent and are not biodegradable; these substances cannot be removed via either coagulation with metal salts or multilayer or biological filtration (Ternes *et al.* 2002; Westerhoff *et al.* 2005; Vieno *et al.* 2006; Bundy *et al.* 2007; Onesios *et al.* 2009). Advanced technologies including high-pressure membrane filtration should thus be considered with respect to the removal of a variety of OMPs.

A broad spectrum of trace organic substances can be treated effectively via the use of powdered (PAC) or granular (GAC) activated carbon adsorption (Choi *et al.* 2008; Grover *et al.* 2011; Serrano *et al.* 2011). Nevertheless, certain single substances are again persistent and cannot be removed by these processes. The presence of NOM leads to competitive adsorption, which may decrease the removal of micropollutants (Sontheimer *et al.* 1985), while the use of oxidants prior to activated carbon decreases the effectiveness of the latter regarding target compounds (Karanfil, 2006). As a result, there is a risk that in the conventional treatment scheme, pre-chlorination, UV-disinfection and ozonation may reduce the adsorption of OMPs on granular activated carbon (Figure 7.5). Additionally, the suitability of different activated carbons should be considered, as they may exhibit varying levels of effectiveness in terms of the removal of different OMPs.

Nanofiltration (NF) and reverse osmosis (RO) are typically considered the most reliable methods of OMP treatment, allowing >70% or even complete removal of most of these trace substances (Xu *et al.* 2005; Snyder *et al.* 2007; Malaeb & Ayoub, 2011). This is due to the fact that NF and RO membranes have apparent molecular weight cut-off values of around 150–200 g/mol, the same size range as that of most trace micropollutants. Nevertheless, membrane performance with respect to OMP reduction can be limited by certain factors, such as pore size (>OMP size), solute-membrane affinity, electrostatic and adsorptive interactions between membrane and solutes, and feed water organic matrix (Bellona *et al.* 2004; Siegrist & Joss 2012). In practice, this implies the problematic rejection of certain neutral, highly hydrophobic compounds (i.e., carbamazepine, bisphenol-A), small positively-charged pollutants (e.g., pharmaceutical terbutaline), as well as extremely small polar pollutants (e.g., NDMA).

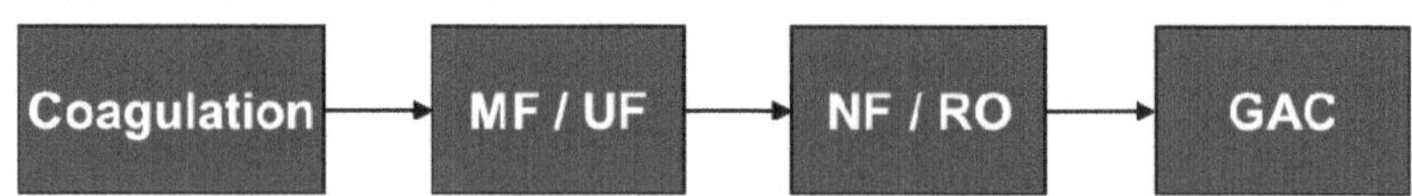

Figure 7.6 Suggested optimised scheme for Lake Paranoá water treatment. MF, UF and NF stand for micro-, ultra- and nanofiltration, respectively, RO for reverse osmosis and GAC for granular activated carbon.

Therefore, polishing with activated carbon adsorption may be needed in addition to NF or RO (Snyder *et al.* 2007; Verliefde *et al.* 2007). One of the reported advantages of this combination of activated carbon adsorption with membrane processes is that it can effectively remove both NOM and trace organic compounds, including disinfection by-products, thus providing a robust dual barrier for the removal of the latter. Whereas the use of NF or RO membranes enables the removal of both high and medium molecular weight NOM, as well as a number of OMPs, GAC more effectively adsorbs non-polar compounds (Baldauf *et al.* 2013; Bethmann *et al.* 2013). Taking the above-mentioned factors into account, the suggested technological alternative regarding the treatment of Lake Paranoá water is membrane (NF or (ultra)-low pressure RO) filtration, with a pre-treatment such as coagulation and ultrafiltration, and a follow-up treatment involving granular activated carbon (Figure 7.6). Water recovery in this case would be expected to be high (up to 90%) due to the constant annual temperature of between 20 and 25°C and mainly low mineralisation of the lake water (TDS ca. 60 mg/L). As a result, the NF/RO membranes would be expected to exhibit higher permeability and lower scaling (Bethmann *et al.* 2013), thus also enabling the environmentally friendly possibility of concentrate discharge directly into the sewage system due to the small concentrate flux and its low salt content.

7.3.3 Additional measures to bridge the gap between water demand and availability

Nevertheless, even if a new treatment plant on Lake Paranoá is put into operation, the capacity of the extended water supply system may not be able to meet the increasing demand in the long term, based on the district's projected population growth rates. In addition, the currently observed dissipating method of water use will continue to widen the gap between water availability and demand. In this respect, the importance of raising public awareness should be underlined for Brasília as an essential component of water conservation in the city, along with the use of innovative technologies. Following the large variety of programmes currently being carried out in other Brazilian regions which are aimed at reducing water wastage, there is an urgent need to increase public participation in the DF in order to guarantee a stable and sufficient water supply in the future. The scheme with the largest potential for the DF involves the optimisation of water use in private households, taking into account the fact that present water consumption ranges tremendously from 70 up to 1000 L per capita per day (personal communication). Water availability per capita could even be increased if savings are also considered in commercial, public and industrial buildings. In some Brazilian regions the average water-saving potential for public buildings has been estimated to be as high as 30% according to the programmes PRO-ÀGUA in the state of São Paulo (Ilha *et al.* 2002; PRO-ÀGUA, 2013) and ÀGUAPURA in the state of Bahia (ÀGUAPURA, 2013). Both programmes are part of the national water conservation project

PURA (PURA, 2013), which has been implemented in a number of regions across the country since the Brazilian Federal Government launched its National Plan against the wastage of water (MPO/SEPURB-PNCDA, 1998) in 1998. Examples from various countries worldwide, such as Germany, the USA and Syria, all of which have successfully conducted public awareness campaigns of this kind in the past, have demonstrated that household potable water savings of up to 25% are possible in the long term, that is, after three or more years (Königs, 1998; Gregg *et al.* 2007; GIZ, 2011).

Last but not least, the DF has tremendous potential for reducing potable water consumption via the use of rainwater. This is especially relevant both for the public sector and residential areas with a lower level of income per family and an inadequate water supply. Whilst global average annual precipitation levels are around 760 mm, in the DF this figure reaches 1550 mm (WMO, 2011). Thus the potential potable water savings which can be made via rainwater usage have been estimated to range between 48 and 100% across the country, with levels in the central-west region which includes the DF of Brasília surpassing 50% (Ghisi, 2006). This perspective is all the more attractive considering that collected rainwater could also have non-potable uses, which usually account for around 50% of water consumption in a household (Ghisi 2006). One such initiative, the 'One million rural cisterns programme'(P1MC), commenced in 2001 in North-East Brazil, a region notable for its semi-arid climate and pronounced water shortages (Gomes *et al.* 2012). Further projects aimed at the promotion of rainwater harvesting should be undertaken in order to prevent water scarcity and avoid critical loading of the water supply system. There is no doubt that such policies which are intended to raise public awareness and participation should become an indispensable element of future IWRM plans in order to ensure sustainable development in the DF.

7.4 CONCLUSIONS

Seasonal variations in rainfall which are typical of its tropical climate play an important role in changing the quality of the surface water sources used for drinking water supply in the DF. Local waterworks thus exhibit variations in removal effectiveness, mainly associated with fluctuations in raw water particulate matter content, which may not only vary by one order of magnitude within a single facility during a single season, but also present even greater variation between wet and dry seasons. Due to their nature, dissolved organic substances (even those of high molecular weight) are not always easy to remove during treatment. Furthermore, it is not easy to achieve the simultaneous optimal removal of both particles and dissolved organic matter.

Based on the findings of the present study, the main recommendations regarding the intended treatment plant on Lake Paranoá in terms of providing high-quality potable water include: (1) additional treatment support via analytical and monitoring methods to deal with natural fluctuations of particles in raw water and (2) the application of advanced treatment technologies to be considered as an alternative to conventional treatment in order to tackle the problem of trace organic substance pollution. Treatment would then include high-pressure membrane filtration (nanofiltration or reverse osmosis) with a conventional train, or better still, clarification combined with ultrafiltration as pre-treatment and adsorption on activated carbon as post-treatment. As a long-term investment, such a decision would also help with meeting the requirements for other drinking water quality parameters, which are becoming increasingly strict worldwide. Finally, it was shown that the DF of Brasília has great potential for water savings both through rationalising the use of available water and through using rainwater to reduce potable water consumption for non-potable purposes.

7.5 ACKNOWLEDGEMENTS

This study was performed within the framework of the German-Brazilian project IWAS-ÁGUA DF, financially supported by the German Federal Ministry of Education and Research (BMBF) via Grants Nr. 02WM1028 and Nr. 02WM1166. The authors gratefully acknowledge the staff at the involved waterworks for their support in providing data, samples and information, Jana Brückner for laboratory assistance, as well as all project partners for successful collaboration throughout the years. The opinions expressed in this paper are exclusively those of the authors and do not necessarily reflect the views of the Brasília water supply and sewerage company.

7.6 REFERENCES

ÁGUAPURA (2013). Programme for the rational use of water and energy (in Portuguese). http://teclim.ufba.br/aguapura/index.php (accessed 21 August 2013).

Altafin I. and Azevedo J. P. (2007). Planned city, sustainable water services? The case of Brasília – the 47-year-old capital of Brazil. In: Proc. of International Symposium on New Directions in Urban Water Management, 12–14 September, UNESCO, Paris.

Amy G. L., Sierka R. A., Bedessen J., Price D. and Tan L. (1992). Molecular size distribution of dissolved organic matter. *J. Am. Water Works Ass.*, **84**, 67–75.

Aster S., Vasyukova E., Ripl K., Kuse B., Günther N., Uhl W., Makeschin F., Wummel J., Krebs P. and Günthert W. F. (2010). Developing an integrated water resources management (IWRM) concept for the capital Brasília. *gwf Wasser-Abwasser*, **151**(special issue), 52–57.

Baldauf G., Brauch H. -J., Haist-Gulde B., Müller U., Sacher F., Rohns H. -P., Konradt N., Wagner C. and Kuhlen J. G. (2013). Activated carbon and reverse osmosis for trace contaminants removal (in German). *Energie – Wasser-Praxis*, **11**, 26–31.

Bedding N. D., McIntyre A. E. and Lester J. N. (1983). Organic contaminants in the aquatic environment III. Public health aspects, quality standards and legislation. *Sci. Total Environ.*, **27**, 163–200.

Bellona C., Drewes J. E., Xua P. and Amy G. (2004). Factors affecting the rejection of organic solutes during NF/RO treatment – a literature review. *Water Res.*, **38**, 2795–2809.

Bethmann D., Baldauf G., Rhode K. and Müller U. (2013). Removal of polar trace substances from a low-mineralized water using membrane technology (in German). *gwf Wasser-Abwasser*, July-August, 844–853.

Briley D. S. and Knappe D. R. U. (2002). Optimizing ferric sulfate coagulation of algae with streaming current measurements. *J. Am. Water Works Ass.*, **94**(2), 80–90.

Bundy M. M., Doucette W. J., McNeill L. and Ericson J. F. (2007). Removal of pharmaceuticals and related compounds by a bench-scale drinking water treatment system. *J. Water Supply: Res. Technol.-AQUA*, **56**(2), 105–115.

CAESB (2004). DF Master Plan for Water and Sanitation (in Portuguese). Government of the DF, Brasilia, Brazil.

Choi K.-J., Kim S.-G. and Kim S.-H. (2008). Removal of tetracycline and sulfonamide classes of antibiotic compound by powdered activated carbon. *Environ. Technol.*, **29**(3), 333–342.

CONAMA (2005). Resolução No. 357, de 17 de março de 2005, Conselho Nacional do Meio Ambiente: Regulation on the quality of surface water resources intended for production of potable water (in Portuguese).

Cotruvo J. A. (1985). Organic micropollutants in drinking water: an overview. *Sci. Total Environ.*, **47**, 7–26.

Dentel S. K., Thomas A. V. and Kingery K. M. (1989). Evaluation of the streaming current detector – i. use in jar tests. *Water Res.*, **23**(4), 413–421.

Fellman J. B., Hood E., Edwards R. T. and D'Arnore D. V. (2009). Changes in the concentration, biodegradability, and fluorescent properties of dissolved organic matter during stormflows in coastal temperate watersheds. *J. Geophys. Res.*, **114**(G1).

Ghisi E. (2006). Potential for potable water savings by using rainwater in the residential sector of Brazil. *Building and Environment*, **41**, 1544–1550.

GIZ (Deutsche Gesellschaft für Internationale Zusammenarbeit) (2011) www.giz.de (accessed 30 August 2011).

Gomes U. A. F., Heller L. and Pena J. L. (2012). A national program for large scale rainwater harvesting: an individual or public responsibility? *Water Resources Management*, **26**(9), 2703–2714.

Gregg T., Strub D. and Gross D. (2007). Water efficiency in Austin, Texas, 1983–2005: an historical perspective. *J. Am. Water Works Ass.*, **99**(2), 76–86.

Gregor J. E., Nokes C. J. and Fenton E. (1997). Optimising natural organic matter removal from low turbidity waters by controlled pH adjustment of aluminium coagulation. *Water Res.*, **31**, 2949–2958.

Grover D. P., Zhou J. L., Frickers P. E. and Readman J. W. (2011). Improved removal of estrogenic and pharmaceutical compounds in sewage effluent by full scale granular activated carbon: Impact on receiving river water. *J. Hazard. Mater.*, **185**(2–3), 1005–1011.

Grünheid S., Amy G. and Jekel M. (2005). Removal of bulk dissolved organic carbon (DOC) and trace organic compounds by bank filtration and artificial recharge. *Water Res.*, **39**(14), 3219–3228.

Hebben N., Abbt-Braun G., Steiniger B., Börnick H., Worch E., Cavalcanti C. G. B., Cavalcanti C. P. and Frimmel F. (2011). Assessment of organic micropollutants, nutrients and dissolved organic carbon in Lake Paranoá – the basis for optimized water treatment. In: Proceedings of the 12th International Conference on Watershed & River Basin Management, Recife, Brazil.

Heberer Th., Schmidt-Bäumler K. and Stan H.-J. (1998). Occurrence and distribution of organic contaminants in the aquatic system in Berlin. Part I: Drug residues and other polar contaminants in Berlin surface and groundwater. *Acta Hydrochim. Hydrobiol.*, **26**, 272–278.

Henderson R. K., Baker A., Parsons S. A. and Jefferson B. (2008). Characterisation of algogenic organic matter extracted from cyanobacteria, green algae and diatoms. *Water Res.*, **42**, 3435–3445.

Huber S. A., Balz A., Abert M. and Pronk W. (2011). Characterisation of aquatic humic and non-humic matter with size-exclusion chromatography – organic carbon detection – organic nitrogen detection (LC-OCD-OND). *Water Res.*, **45**, 879–885.

Hurst A. M., Edwards M. J., Chipps M., Jefferson B. and Parsons S. A. (2004). The impact of rainstorm events on coagulation and clarifier performance in potable water treatment. *Sci. Total Environ.*, **321**, 219–230.

Ilha M. S. O., Nunes S. S. and Pedroso L. P. (2002). Water conservation program in the State University of Campinas. In: Proceedings of the 28th International Symposium on Water Supply and Drainage for Buildings, September 2002, Iasi, Romania.

Jardé E., Gruau G. and Mansuy-Huault L. (2007). Detection of manure derived organic compounds in rivers draining agricultural areas of intensive manure spreading. *Appl. Geochem.*, **22**(8), 1814–1824.

Kalbus E., Kalbacher T., Kolditz O., Krüger E., Seegert J., Teutsch G., Borchardt D. and Krebs P. (2012). IWAS – Integrated Water Resources Management under different hydrological, climatic and socio-economic conditions. *Environ. Earth Sci.*, **65**, 1363–1366.

Karanfil T. (2006). Chapter 7. Activated carbon adsorption in drinking water treatment. In: Interface Science and Technology. J. B. Teresa, Elsevier. Volume **7**, 345–373.

Königs T. (1998). Das Wasserspar-Buch (Water-saving book; in German). Falken Verlag, Würzburg.

Lee G.-s., Lee K.-h. and Jeong G.-c. (2009). A strategy for quantifying turbid water occurrence possibility based on geologic characteristics and soil erosion in hydrologic basins. *Environ. Earth Sci.*, **59**, 821–835.

Loraine G. A. and Pettigrove M. E. (2006). Seasonal variations in concentrations of pharmaceuticals and personal care products in drinking water and reclaimed wastewater in southern California. *Environ Sci. Technol.*, **40**(3), 687–695.

Lorz C., Abbt-Braun G., Bakker F., Borges P., Börnick H., Fortes L., Frimmel F. H., Gaffron A., Hebben N., Höfer R., Makeschin F., Neder K., Roig H. L., Steiniger B., Strauch M., Walde D. H., Weiß H., Worch E. and Wummel J. (2012). Challenges of an integrated water resource management for the Distrito Federal, Western Central Brazil: climate, land-use and water resources. *Environ. Earth Sci.*, **65**, 1575–1586.

Maeng S. K., Salinas Rodriguez C. N. A. and Sharma S. K. (2013). Chapter 13. Removal of pharmaceuticals by bank filtration and artificial recharge and recovery. In: Comprehensive analytical chemistry. D. B. Mira Petrovic and P. Sandra (eds.), Elsevier. Volume **62**, 435–451.

Maier H. R., Morgan N. M. and Chow Ch. W. K. (2004). Use of artificial neural networks for predicting optimal alum doses and treated water quality parameters. *Environmental Modelling and Software*, **19**(5), 485–494.

Malaeb L. and Ayoub G. M. (2011). Reverse osmosis technology for water treatment: State of the art review. *Desalination*, **267**(1), 1–8.

Mälzer H. -J. and Strugholz S. (2008). Artificial neural networks for cost optimization of coagulation, sedimentation and filtration in drinking water treatment. *Water Sci. Technol.: Water Supply*, **8**(4), 383–388.

Mania M. (2006). Influence of algal matter on coagulation and adsorption processes in water treatment (in German). PhD thesis, Technische Universität Berlin, pp. 141.

Mierzwa J. C., Costa Cabral da Silva M., Rodrigues Valadares Veras L., Lucas Subtil E., Rodrigues R., Li T. and Raquel Landenberger K. (2012). Enhancing spiral-wound ultrafiltration performance for direct drinking water treatment through operational procedures improvement: A feasible option for the Sao Paulo Metropolitan Region. *Desalination*, **307**, 68–75.

Morel B., Durand P., Jaffrezic A., Gruau G. and Molenat J. (2009). Sources of dissolved organic carbon during stormflow in a headwater agricultural catchment. *Hydrol. Process.*, **23**, 2888–2901.

Nam S. -W., Jo B. -I., Kim M. -K., Kim W. -K. and Zoh K. -D. (2013). Streaming current titration for coagulation of high turbidity water. *Colloids Surf., A*, **419**, 133–139.

Onesios K. M., Yu J. T. and Bouwer E. J. (2009). Biodegradation and removal of pharmaceuticals and personal care products in treatment systems: a review. *Biodegradation*, **20**(4), 441–466.

Portaria (2004). Norma de qualidade da àgua para consumo humano. Regulation on the quality of water intended for human consumption (in Portuguese). Portaria MS Nr. 518/2004

POSEIDON (2004). Assessment of technologies for the removal of pharmaceuticals and personal care products in sewage and drinking water facilities to improve the indirect potable water reuse (POSEIDON), detailed project report. http://www.eu-poseidon.com.

PRO-ÁGUA (2013). Water conservation programme in the UNICAMP (in Portuguese). http://www.fec.unicamp.br/~milha/proagua.htm (accessed 21 August 2013).

PURA (2013). Programme for the rational use of water (in Portuguese), Sabesp. http://www.sabesp.com.br (accessed 29 June 2013).

MPO/SEPURB-PNCDA (1998). Ministério do Planejamento e Orçamento – Secretaria de Política Urbana. National programme to combat water waste 'Programa Nacional de Combate ao Desperdício de Água, PNCDA'(in Portuguese). Estrutura do Programa - Documento Técnico de Apoio 1 (DTA-1), p. 31.

Reungoat J., Escher B. I., Macova M. and Keller J. (2011). Biofiltration of wastewater treatment plant effluent: effective removal of pharmaceuticals and personal care products and reduction of toxicity. *Water Res.*, **45**(9), 2751–2762.

Rigobello E. S., Dantas A. D. B., Di Bernardo L. and Vieira E. M. (2013). Removal of diclofenac by conventional drinking water treatment processes and granular activated carbon filtration. *Chemosphere*, **92**(2), 184–191.

Semmens M. J. and Field T. K. (1980). Coagulation: experiences on organics removal. *J. Am. Water Works Ass.*, **72**, 476–483.

Serrano D., Suárez S., Lema J. M. and Omil F. (2011). Removal of persistent pharmaceutical micropollutants from sewage by addition of PAC in a sequential membrane bioreactor. *Water Res.*, **45**(16), 5323–5333.

Siegrist H. and Joss A. (2012). Review on the fate of organic micropollutants in wastewater treatment and water reuse with membranes. *Water Sci. Technol.*, **66**(6), 1369–1376.

Singer P. C. (1999). Humic substances as precursors for potentially harmful disinfection by-products. *Water Sci. Technol.*, **40**(9), 25–30.

Snyder S. A., Adham A., Redding A. M., Cannon F. S., DeCarolis J., Oppenheimer J., Wert E. C. and Yoon Y. (2007). Role of membranes and activated carbon in the removal of endocrine disruptors and pharmaceuticals. *Desalination*, **202**(1–3), 156–181.

Sontheimer H., Frick B. R., Fettig J., Hörner G., Hubele C. and Zimmer G. (1985). Adsorption processes for water treatment (in German). DGVW-Forschungsstelle am Engler-Bunte-Institut der Universität Karlsruhe, Karlsruhe.

Staaks Ch., Fabris R., Lowe T., Chow Ch. W. K., van Leeuwen J. A. and Drikas M. (2011). Coagulation assessment and optimisation with a photometric dispersion analyser and organic characterisation for natural organic matter removal performance. *Chem. Eng. J.*, **168**(2), 629–634.

Stackelberg P. E., Furlong E. T., Meyer M. T., Zaugg S. D., Henderson A. K. and Reissman D. B. (2004). Persistence of pharmaceutical compounds and other organic wastewater contaminants in a conventional drinking-water-treatment plant. *Sci. Total Environ.*, **329**(1–3), 99–113.

Stackelberg P. E., Gibs J., Furlong E. T., Meyer M. T., Zaugg S. D. and Lee Lippincott R. (2007). Efficiency of conventional drinking-water-treatment processes in removal of pharmaceuticals and other organic compounds. *Sci. Total Environ.*, **377**(2–3), 255–272.

Steiniger B., Börnick H., Hebben N., Abbt-Braun G., Frimmel F. H., Nödler K., Licha T., Cavalcanti C. G. B., Cavalcanti C. P., Brandão C. C. S. and Worch E. (2011). Investigations on the Distribution of Emerging Organic Pollutants in Lake Paranoá – Contributions to the Management of the Lake. In: Proceedings of the 12th International Conference on Watershed & River Basin Management, Recife, Brazil.

Strauch M., Lima J. E. F. W., Volk M., Lorz C. and Makeschin F. (2013). The impact of best management practices on simulated streamflow and sediment load in a Central Brazilian catchment. *J. Environ. Manage.*, **127**, 24–36.

Ternes T. A., Meisenheimer M., McDowell D., Sacher F., Brauch H.-J., Haist-Gulde B., Preuss G., Wilme U. and Zulei-Seibert N. (2002). Removal of pharmaceuticals during drinking water treatment. *Environ. Sci. Technol.*, **36**(17), 3855–3863.

Vasyukova E., Uhl W., Braga F., Simões C., Baylão T. and Neder K. (2012). Drinking water production from surface water sources in the tropics: Brasília DF, Brazil. *Environ. Earth Sci.*, **65**(5), 1587–1599.

Verliefde A. R. D., Heijman S. G. J., Cornelissen E. R., Amy G., Van der Bruggen B. and van Dijk J. C. (2007). Influence of electrostatic interactions on the rejection with NF and assessment of the removal efficiency during NF/GAC treatment of pharmaceutically active compounds in surface water. *Water Res.*, **41**(15), 3227–3240.

Vieno N., Tuhkanen T. and Kronberg L. (2006). Removal of pharmaceuticals in drinking water treatment: effect of chemical coagulation. *Environ. Technol.*, **27**(2), 183–192.

Vu B., Chen M., Crawford R. J. and Ivanova E. P. (2009). Bacterial extracellular polysaccharides involved in biofilm formation. *Molecules*, **14**, 2535–2554.

Westerhoff P., Yoon Y., Snyder Sh. and Wert E. (2005). Fate of endocrine-disruptor, pharmaceutical, and personal care product chemicals during simulated drinking water treatment processes. *Environ. Sci. Technol.*, **39**(17), 6649–6663.

WMO (World Meteorological Organization) (2011). World Weather Information Service: Brasília. http://www.worldweather.org/136/c00290.htm (accessed on May 3, 2011).

Wu G.-D. and Lo Sh.-L. (2008). Predicting real-time coagulant dosage in water treatment by artificial neural networks and adaptive network-based fuzzy inference system. *Eng. Appl. Artif. Intel.*, **21**(8), 1189–1195.

Xu P., Drewes J. E., Bellona C., Amy G., Kim T. -U., Adam M. and Heberer T. (2005). Rejection of emerging organic micropollutants in nanofiltration-reverse osmosis membrane applications. *Water Environ. Res.*, **77**(1), 40–48.

Chapter 8

Developing the urban water system towards using the Paranoá Lake in Brasília as receptor and water resource

F. W. Günthert, V. Freitas, K. Neder, A. Obermayer, S. Faltermaier and C. Tocha

8.1 INTRODUCTION

Reservoir water is one of the most important fresh water resources for the water supply of drinking water in the city of Brasília. Pollution, uncontrolled and strong stormwater runoff, sewer leakage, wastewater discharge, etc. influence the water quality and the aquatic environment in the DF. One possible solution to this challenge is to implement integrated water management practices, including efficient sewage and drainage systems and wastewater treatment (Aster at al. 2010). The generation of an integrated urban water system with inclusion of stormwater management forms the base for sustainable water protection of the Paranoá Lake.

Therefore this topic was the focus of a project group[1] consisting of scientists from the University of the Federal Armed Forces Munich (UniBwM) and Brazilian partners – the CAESB (Water supply and wastewater disposal company of Brasília), the NOVACAP (Company responsible for the drainage system in Brasília) and the UnB (University of Brasília).

8.1.1 Background and aims of urban water management

Urban water management is a technical discipline that deals with all aspects of water in connection with settlements (Gujer, 2008). It has a major importance for human comfort and safety. In addition to the management of natural water resources and supply of safe drinking water, urban water management ensures the discharge of potentially polluted water (wastewater, stormwater) and the treatment and recirculation in the water cycle.

According to the German Association for Water, Wastewater and Waste (DWA, 2006), Figure 8.1 shows the principals of an integrated urban drainage and the main subjects and objects of protection. It is divided into four areas, safe disposal in urban areas, water pollution control, safe usage and other concerns like nature and soil conservation. Due to regional conditions of the study area, the subjects 'drinking water abstraction' and 'flood protection' connected with the issue of 'soil protection' have a particular relevance for Brasília. Therefore the focus of this chapter will put emphasis on these aspects.

To be able to deal with those objectives professionally, a general first view of the system of urban drainage must take place (Figure 8.2). The sources of water that enter the urban system are precipitation as well as domestic and industrial wastewater. Wastewater must be treated without exception. The necessity of rainwater treatment depends on the pollution degree. Uncontaminated water can infiltrate directly through the soil into groundwater or be discharged into the surface water.

[1]The work of this project group and the pilot plant were funded by the BMBF – the German Federal Ministry of Research and Education.

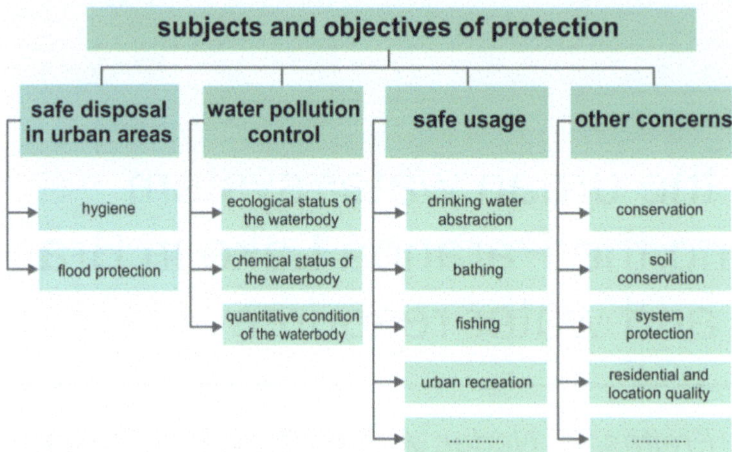

Figure 8.1 Urban drainage: Subjects and objectives of protection, based on DWA (2006).

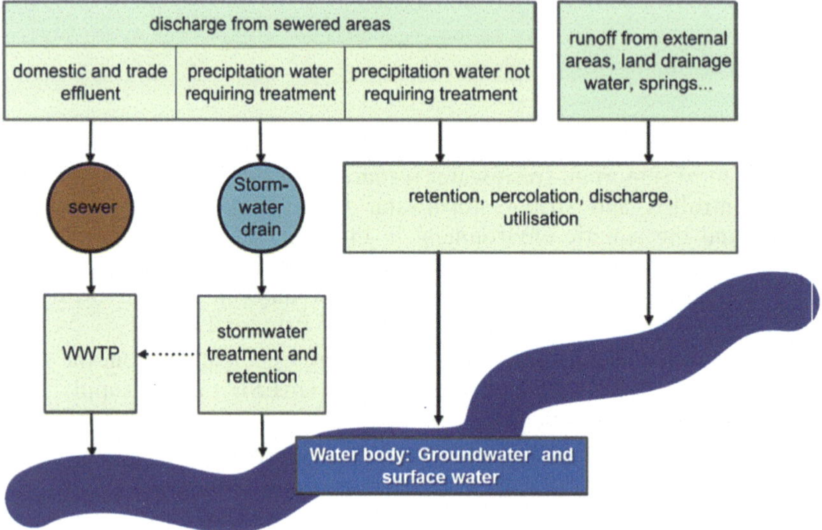

Figure 8.2 The drainage system: Discharge into the receiving water body.

For achieving a sustainable protection of the water body against contamination, it is necessary to look in detail at the way wastewater is discharged and treated. For this purpose, the condition assessment of the sewer and drainage system and the wastewater treatment technique are particularly important factors and will be addressed by the example of Brasília.

8.1.2 Situation in Brasília

Studies performed at the CAESB (local Water Supply and Sewage Company of Brasília) show that there is a need to double the water production capacity of the system by 2030. Ideas to realize this are to target new sources of water supplies located in the DF and surrounding areas. One possible solution to secure the water supply for future generations in the DF is to use water resources from receiving water bodies such as the Paranoá Lake, an artificial lake that stretches around the eastern edge of the city of Brasília. However, at the moment the majority of Brasília´s sewage is treated in the wastewater treatment plants ETE Sul and ETE Nortè, which are discharging into the Paranoá Lake. Prior to further planning of a possible to use the lake as a resource for drinking water, a high water quality in the future needs to be secured. In contrast, drinking water supply and wastewater disposal in Brasília are faced with significant challenges. Pollution from stormwater runoff, sewage leakage and wastewater discharge, etc., are major problems to more than just the aquatic environment. Thus, the project is focused on the Paranoá Lake and its catchment area. General aims are to improve the quality of receiving waters (Paranoá as a water supply source) by optimizing sewers and drainage network management.

Due to the climatic conditions in Brasília and the seasonal change between dry season and rainy season there are regular rainfall events that exceed the capacity of the drainage system. This leads to an increase of flooding and erosion. In addition to the loss of soil, erosion might cause as off-site effect the deterioration of water quality by increased turbidity. These are serious problems with regard to the use of the lake as a drinking water source.

Brasília´s urban drainage system is designed in a conventional concept which consists of collection, inlet, network and final discharge into the Paranoá Lake. Physical space limitations around the center of the city have not yet impeded the implementation of stormwater treatment plants. In contrast to this, untreated urban runoff is one of the main reasons for water pollution. There are around 100 discharging points from galleries in the Paranoá Lake basin through which for example, suspended solids and organic and inorganic compounds enter directly into the Paranoá Lake. Moreover, illegal connections from sewers lead to an additional load of for example, coliforms from discharging points. The aim according to Integrated Water Resources Management (IWRM) is to either not let these substances enter the water path at all or only let them enter the water in little concentrations. As a first step, in order to estimate the pollution load from discharging points and derive possible issues, the discharging points have to be monitored. Figure 8.3 shows elements of the existing drainage system and one of the discharging points in Brasília.

Figure 8.3 Pictures above: elements of the discharge system in Brasília; picture bottom left: sedimentation basin; picture bottom right: discharging point.

An earlier monitoring of the 16 largest discharge points was done by CAESB in the period from 2003 to 2006 and provided the first indications of the discharge points with the highest pollution load (Figure 8.4). These points should be incorporated into the new monitoring and furthermore to the catchment areas with a high number of illegal connections.

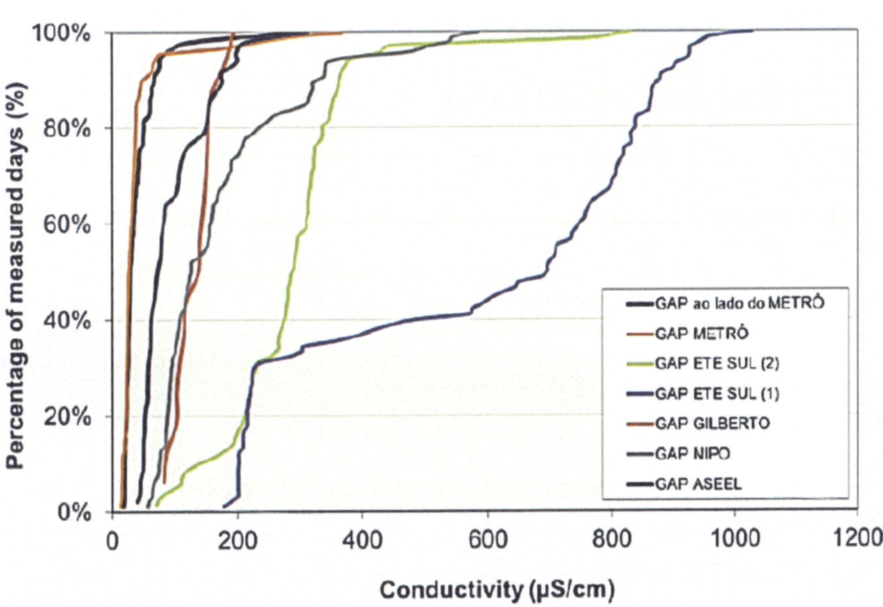

Figure 8.4 Results of conductivity monitoring as indicator for pollution load (GAP = Stormwater discharge point).

Brasília has a separate system for sewage and rainwater collection. The state owned company NOVACAP operates the urban drainage system, while the sewage system and wastewater treatment plants are run by the state owned water supply and wastewater company CAESB. For sewers the specific renovation rate is 0.6% per year; for drains there is no data available. Due to the fact that Brasília was build nearly 50 years ago, one can figure out that great parts of the sewer and drain system are of the same age. Therefore, a likely threat are damages in the channels that may lead to exfiltration of wastewater and infiltration of rainwater. This in turn results in an increased hydraulic pressure on wastewater treatment plants and unwanted dilution of the wastewater that needs to be treated. As a result, the cleaning performance of the wastewater treatment plant will be reduced and the potential of polluted water reaching the lake increases.

A further critical aspect is the existing wastewater treatment plant technology has limitations in terms of trace substances or so called micropollutants/-substances. Over the last few years there has been a growing concern over the increasing concentration of microsubstances originating from a great variety of sources including pharmaceutical, chemical engineering and personal care products in rivers, lakes, soil and groundwater (Ternes & Joss, 2006; DWA, 2008; Abegglen, 2009; Abegglen & Siegrist, 2012). These substances enter the water cycle mainly through industrial or municipal discharge or wastewater from hospitals (Track & Kreysa, 2003; Kloepfer, 2005; Reemtsma & Jekel, 2006; Bester, 2008). The biggest part of the trace substances are polar and persistent compounds that can only partially or not at all be removed from wastewater with conventional wastewater treatment and thus can enter the environment. And for most of these substances the hazardous potential for biological systems still has an unknown extent (see Chapter 5 of this book).

In summary, the following approaches were planned to deal with the challenge to protect the Paranoá Lake from pollutants entering through precipitation and wastewater and further developing the urban water system towards using the Paranoá Lake in Brasília as receptor and water resource,

- Analysis of the sewer and stormwater system
- Benchmarking for sewers and the wastewater treatment plants
- Implementation of a new monitoring program for stormwater discharging points
- Proposal and development of a concept for efficient sewer management
- Identification of illegal connections to drain and sewers
- Development of a method for studying sewer infiltration and exfiltration
- Identify new wastewater treatment techniques for Brasília and implement a pilot plant for advanced wastewater treatment
- Development and proposal for achieving a sustainable stormwater management
- Regarding energy aspects in the wastewater treatment plants and the pumping stations
- Ascertainment of a concept for efficient sludge management

We selected two approaches assumed to be the most powerful to achieve the targeted goals for Brasília, that is, advanced wastewater treatment and sustainable stormwater management.

8.2 ADVANCED WASTEWATER TREATMENT

Organic microsubstances are typically detectable in microgram to nanogram concentrations. Despite well-established technology it is still not possible to completely remove many of the substances from the wastewater (Abegglen & Siegrist, 2012). The effect of residues of microsubstances on the human body is for the most part still unknown (BLAC, 2003). However, influence on the fertility of various animals due to endocrine properties has been proven (Trachsel, 2008). Therefore, water resources need to be protected from these kinds of inputs, especially in the case of the Paranoá Lake, which is intended to be used as a freshwater reserve in the future. Reduction of input can be achieved by avoidance strategy, decentralized treatment of highly concentrated wastewater, or 'end-of-pipe' approach, which focuses on improving existing treatment plants towards higher elimination of microsubstances. Investigations showed that only membrane filtration, sorption- and oxidation techniques work as 'end-of-pipe' technologies.

8.2.1 Overview of advanced wastewater treatment technologies

8.2.1.1 Biodegradation and sorption in the activated sludge process

Biodegradation depends on a multitude of influences, such as the presence of the specific substance, milieu conditions, diversity of organisms in the systems and overall process conditions (aerobic, anaerobic). These conditions change during biological degradation. Especially sludge age and composition of the biocenosis strongly influence the elimination performance. Significantly increased elimination was observed for higher sludge ages (Ternes & Joss, 2006; Fahlenkamp *et al.* 2006; Ivashechkin 2006; Cornel & Meda, 2009). Highly diversified biocenoses lead to better results accordingly. Chemical reactions in the aqueous phase are only of minor importance for degradation of microsubstances.

8.2.1.2 Membrane technology

Membrane technology has successfully been used for decades for the treatment of landfills, leachate and drinking water. Further developments of low pressure submerged membranes lead to the use of membrane bioreactors (MBR) also for municipal wastewater. Direct retention of microsubstances can only be expected with nanofiltration or reverse osmosis systems. However, currently these cannot be operated cost effectively because of very high energy consumption (Knepper, 2004; Friedrich, 2005).

8.2.1.3 Activated carbon adsorption

Originally, activated carbon (AC) adsorption was used in drinking water treatment, mainly for the elimination of organic and biological matter as well as non-degradable substances. The principle is based on the adsorption of mainly non-polar substances to the very high specific surface area of activated carbon of 500–1200 m /g. AC can be used in granular or powdered form. For the application of powdered activated carbon it has to be considered, that competitive reactions of different organic substances may lead to the release of already adsorbed pollutants (Schrader, 2007). This happens especially when AC is directly mixed in the activated sludge tank. When AC is applied behind the biological treatment step, the concentration of organic substances is lower and therefore this effect can be reduced (Cornel, 2010). The AC is then flocked by flocculation agents and separated from the treated wastewater by sedimentation or filtration. In the end the activated carbon is burned, which leads to the mineralization of the organic loading.

An alternative to the use of powdered AC (applied in the biological treatment step, or effluent of secondary clarifier) can be the use of granulated AC, which can be used in fixed-bed adsorbers or GAC filters. The stationary AC layer is hereby flown through by the loaded water (mostly downstream). Precondition for good operation is pretreatment in order to eliminate suspended solids, which may clog the filter. Additional backwashing can remove suspended solids which may still be caught by the AC filter. Experiences from drinking water treatment show that filtration with granular activated carbon (GAC) is currently the most efficient way to remove pharmaceuticals of very low concentrations. Also the overall effluent quality shows significantly lower COD- and BOD values (Cornel & Meda, 2009). Costs for new tanks, the production of AC and an increase of excess sludge of about 10% have to be calculated (Ternes *et al.* 2008; Schrader, 2007; Abegglen, 2009).

8.2.1.4 Advanced oxidation technologies

Advanced oxidation, such as ozonation, has been successfully used in drinking water treatment for a long time for the removal of organic matter and reduction of bacteria (e.g., coliforms). Hardly degradable pharmaceutical such as diclofenac, clarithromycin, metoprolol and even carbamazepine can be eliminated by almost 100%, while the x-ray agent iopromid is barely removed (Abegglen, 2009). This technique, however, does not completely oxidize microsubstances but creates transformation products which may be even more dangerous than the original substance (Cornel & Meda, 2009). Costs of ozonation amount to 0.05 to 0.15 €/m (Joss *et al.* 2008).

8.2.2 Conception, invitation of tenders and delivery of the pilot plant

At the end of 2010 the pilot plant setup was conceived, based on literature research and discussions with the Brazilian partners and was re-engineered at the beginning of 2011. For the concept of the WWTP ETE Nortè a pilot plant with ultrafiltration (UF) in combination with inline-flocculation for the elimination of turbidity and reduction of diluted organic carbon (DOC), as well as downstream AC filter for the retention of organic microsubstances was designed and built by BAUER Umwelt GmbH (Schrobenhausen, Germany).

8.2.3 Process design of the pilot plant

The plant for advanced wastewater treatment is mainly based on the UF units for the reduction of residual particulate matter and diluted organic carbon. The process design of the pilot plant can be seen in Figure 8.5. Raw water is being pumped into the raw water tank by a pressure controlled raw water pump. The process pumps then transports the raw water through the sandfilter, along the flocculation agent dosing station, through the ultrafiltration and activated carbon filter into the clear water tank. The water flow can be adjusted by an inclined-seat valve and a flow meter.

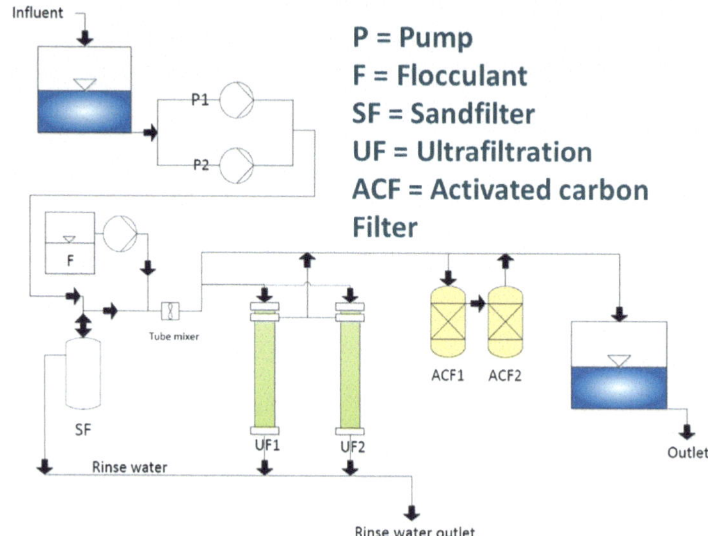

Figure 8.5 Process design of the pilot plant.

The sandfilter serves as prefiltration of the treatment plant with a volume of 30 liters. Filtration material is filter granulate by Turbidex (Harlingen, USA), an Aluminum-Silicate-Mineral. Within the sandfilter all turbid materials >25 μm can be removed from the raw water. This pretreatment was introduced in order to prevent overloading of the ultrafiltration step, as effluent analysis of ETE Nortè was not available at this time. Between sandfilter and ultrafiltration, in-line flocculation has been installed to minimize biofouling by adsorption of organic material to the UF modules. The retained flocs are removed from the membrane by backwashing. The flocculation agent $FeCl_3$ or Aluminiumsulfate is dosed by a dosing pump into the in-line mixer. A sufficient pipelength of 4 m supports the generation of flocs.

The UF unit consists of 2 parallel operated modules (4"UF-elements with Multibore 1.5) by inge watertechnologies GmbH (Greifenberg, Germany) with a total membrane surface of 8 m . These modules can either be operated in dead-end or cross-flow mode. Waterflow necessary for crossflow can be adjusted by a ball valve and a flow meter. The UF serves as a further removal step of turbid materials and reduction of DOC in combination with flocculation. This is important to prevent competitive reactions between microsubstances and natural organic matter in the following activated carbon filter. The adsorption of organic microsubstances is achieved by two either parallel or in series operated AC filters with a combined volume of 150 l. To prevent recontamination, silver was added to the AC.

8.2.4 Flushing program

The efficient operation of filtration systems is very much influenced by the flushing of their components. Flushing of the sandfilter is necessary to loosen the material and remove the filtercake, which also reduces the pressure difference within the filter. Filtration, backwash and rinsing of the SF work fully automatic. Three modes for cleaning of the UF modules are available,

- Forward flush: Cleans the membrane surface from the inside of the individual capillaries
- Backwash: Each UF module is flushed with the permeate of the other module, thus preventing contamination of the modules on the permeate side. This is sufficient for low turbidity wastewater, while at high turbidity the backwash can be supported by adding the membrane pressure tank. The membrane pressure tank will be filled immediately before the backwash with permeate, until it is under a pressure of 5 bar. After opening the pressure valve a high volume flow flushes the UF module from the permeate side. The adhering particles on the raw water side are replaced by this.
- Chemical backwash: For disinfection at pH 12.5 and removal of very persistent contamination (limestone) at pH 2.5. This rinsing water is collected in a neutralization tank and then released fully automatic into the flushing pipe.

8.2.5 Testing phase WWTP Holzkirchen, Germany

The pilot plant was tested in Germany. The WWTP Holzkirchen (30.000 PE) seemed reasonable for this process due to the following boundary conditions:

- Stable operation of the different treatment steps
- Mostly drainage in separating system (like in Brasília)

- Hardly any infiltration water
- Close contact between UniBwM and staff at WWTP Holzkirchen

The tests mainly aimed at safe operation of the different components. It was only operated with dead-end filtration. Flocculation was stopped for a long period of time due to clogging of the magnet valves after backwashing. The pilot plant was operated with the effluent of the secondary clarifier. 16 measuring campaigns were conducted from July-December 2011, evaluating the influent of the pilot plant, the performance of the sandfilter, UF modules and AC filtration. The samples were each taken within one filtration cycle. Conductivity, turbidity and pH were measured directly at the plant, the other values such as TOC reduction (Figure 8.6) in the laboratory of the UniBwM.

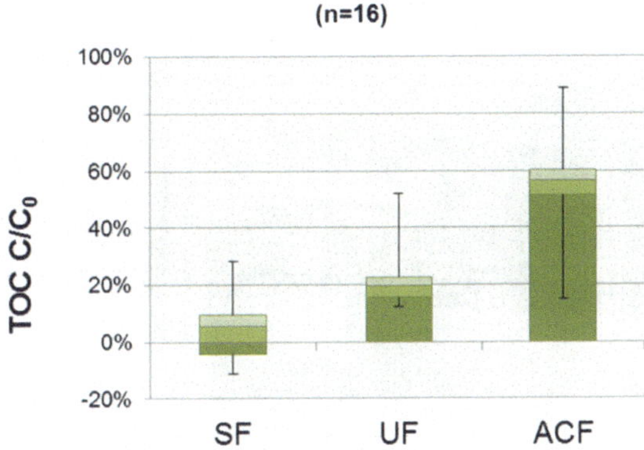

Figure 8.6 Average effluent values – TOC reduction: SF – sandfilter, UF – ultrafiltration, ACF – AC filtration.

Due to technical difficulties the plant was not in operation for longer time, thus causing the risk of uncontrolled bacterial growth in the ACF. These difficulties also led to strong membrane fouling, which resulted in TMP (transmembrane pressures) >3 bar at maximum. Figure 8.7 shows how increasing the flux from 25 l/(m * h) to 50 l/(m * h) directly influences the TMP. It was therefore considered unnecessary to test for organic microsubstances, as boundary conditions may have lead to wrong results. Experiences from WWTP Holzkirchen showed that the pilot plant could be operated effectively at a Flux <25 l/(m * h), filtration intervals of 20 minutes and a CEB (chemical backwash) interval of 24 hours. A higher flux immediately led to a loss of performance, at least without in-line flocculation. For operation in Brasìlia, the cleaning strategy and flux had to be adapted to the local raw water quality and the effect of in-line flocculation had to be tested.

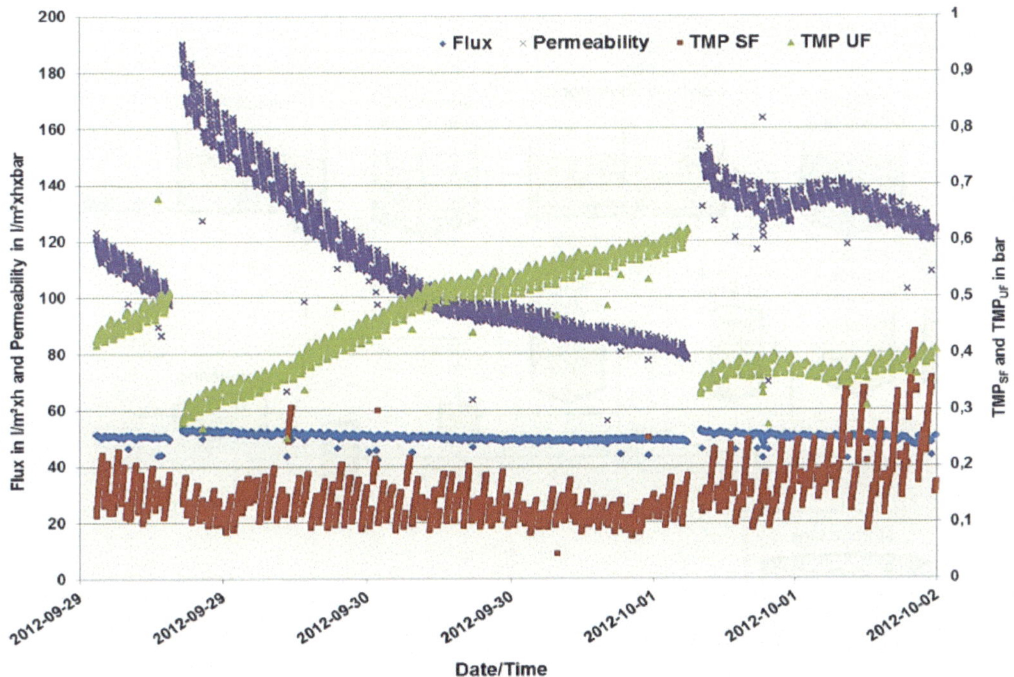

Figure 8.7 Influence of increased flux and CEB on TMP and permeability.

8.2.6 Testing phase at WWTP ETE Nortè

Possible sites for the pilot plant in Brazília were either ETE Nortè or ETE Sul. ETE Nortè was chosen, because of a more stable operation according to CaesB and because of its close vicinity to UnB, which gave scientific and operational support during the project. A picture of the pilot plant are displayed in Figure 8.8.

Figure 8.8 The pilot plant at ETE Nortè in Brasília.

ETE Nortè has been in operation since 1994 with a design capacity of 260,000 PE. The different treatment steps of the WWTP can be seen in Figure 8.9. The pilot plant was set up close to the flotation step and put into operation in March 2012. Damages during the shipping and technology which was not adapted to the Brazilian power supply system delayed stable and continuous operation until September 2012.

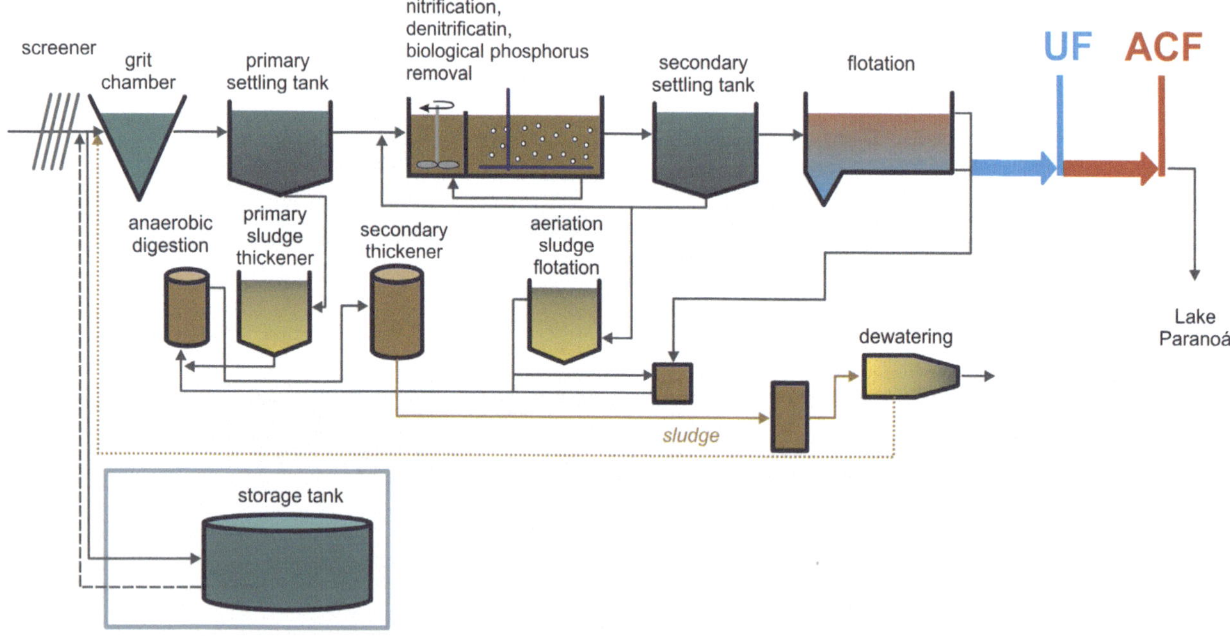

Figure 8.9 Treatment steps of ETE Nortè and position of the pilot plant.

For operation, raw water was taken from the flotation effluent. This effluent already had a high share of flocks, which is why the in-line flocculation of the pilot plant was not used as it had no considerable effect. For the operation of the pilot plant, data from April–June and September–October 2012 is given. A stable operation could be achieved after optimization. When the samples for microsubstances were taken, the plant was operated under the conditions described in Table 8.1.

Table 8.1 Operation conditions of the pilot plant.

Settings				
Flux	l/(m² × h)	50		
Flush interval	Min	25		
Rinse speed	l/(m² × h)	250 (duration 45 – 60 s)		
Rinsing after backwashing	l/(m² × h)	250 (duration 30 s)		
Resulting operational data				
		mean	min	max
Permeability	l/(m² × h)	117	56	208
Transmembrane pressure UF	bar	0.44	0.27	0.62
Inlet pressure sandfilter	bar	0.15	0.05	0.44

Chemically supported membrane rinsing was conducted daily at pH 12.5 and 2.5. Contact time was 15 minutes for each pH-value. To investigate the performance of removal of microsubstances, 24 hours composite samples were analysed for 29 substances, of which only 6 were detectable. The analyses of the water samples were carried out by KIT (Karlsruhe Institute for Technology). Selection criteria were bad biodegradability and ecotoxicological effects, which is the case for example for carbamazepine, atrazine, x-ray agent, ibuprofen or diclofenac (LANUV NRW, 2007).

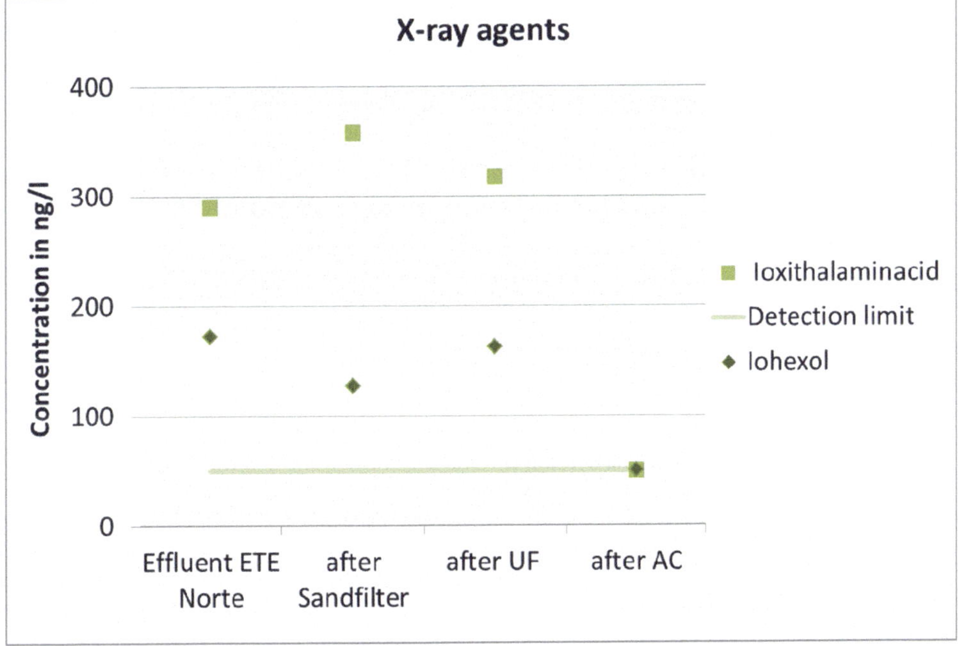

Figure 8.10 Stable operation of the pilot plant: Exemplary removal data for x-ray agents; sample analysis by KIT (Karlsruhe Institute for Technology).

For all groups of substances the same behavior could be observed. Neither sandfiltration nor ultrafiltration had a considerable effect on the concentration of microsubstances. After filtration with granulated activated carbon, however, all investigated substances could be removed below the detection limit (Figure 8.10). Overall stable operation could be achieved on site at ETE Nortè. Several filtration techniques are necessary for the raw effluent from the flotation step in order to protect the consecutive treatment steps. In the end it could be shown that with the use of this pilot plant setup the input of microsubstances into the Paranoá Lake from the wastewater treatment plant ETE Nortè can be reduced considerably. In concsequence, potential sources for future pollution might be reduced to diffuse paths, that is, polluted groundwater.

8.3 STORMWATER MANAGEMENT

The management of stormwater needs to be adapted to the regional and hydrological conditions. One of the key factors for the current discussion about stormwater management is the future increase of runoff in settlements. This development is driven by a worldwide tendency for urbanization and the change of precipitation in consequence of climate change. Therefore we have to reconsider our actual measures of stormwater management to fulfill the objectives of stormwater management in the future (Figure 8.11).

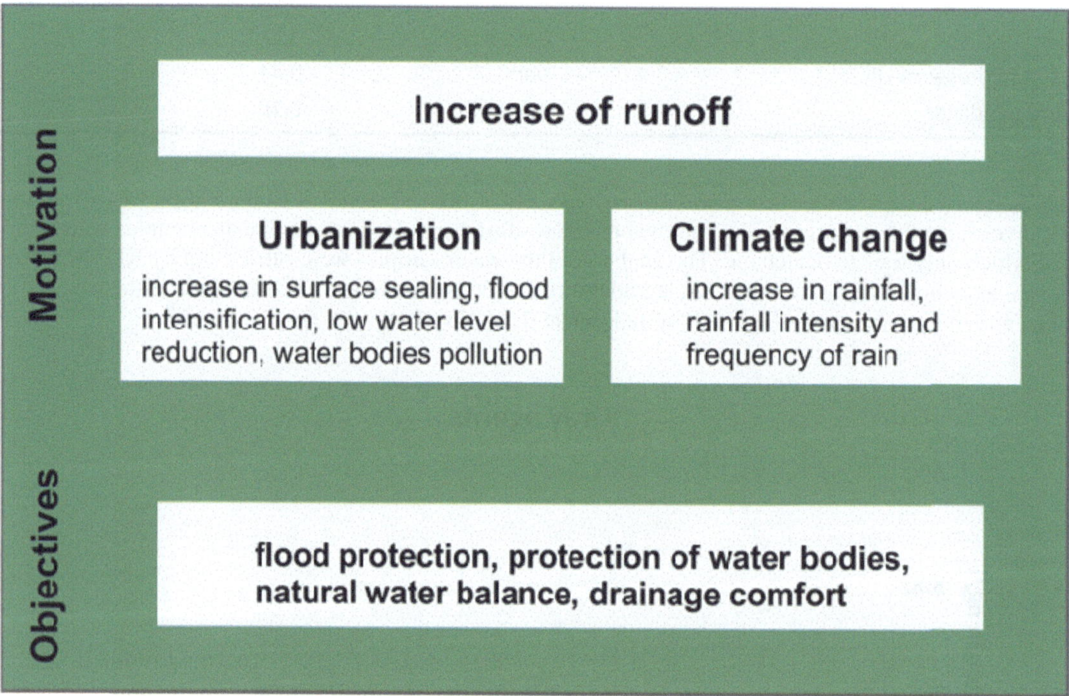

Figure 8.11 Motivation and objectives of stormwater management.

8.3.1 Elements of stormwater management

Prior to decisions on stormwater monagement, a sufficient information base about the quality of stormwater and runoff and it causes has to be created. Only if the stromwater and the runoff fulfill the hygienic and chemical basic requirements or only if these can be reached securely by pretreatment, methods of stormwater management can be applied. Otherwise, stormwater and runoff have to be collected and transported to treatment points or wastewater treatment plants before discharged into water bodies.

There are a number of major elements of stormwater management, from utilization of the stormwater for private or industrial purposes to the centralized retention by storing it. All these elements have a more or less high impact on the natural hydrological budget. They might be ranked by their degree of potential impact into a hierarchy of preference. The major elements are (a) utilization of stormwater; (b) percolation; (c) decentralized retention; (d) centralized retention and (e) extensive drainage.

In most cases it will be not sufficient to use only one of these elements, but a combination is reasonable. In Figure 8.12 a decision support diagram for the selection of the elements is given.

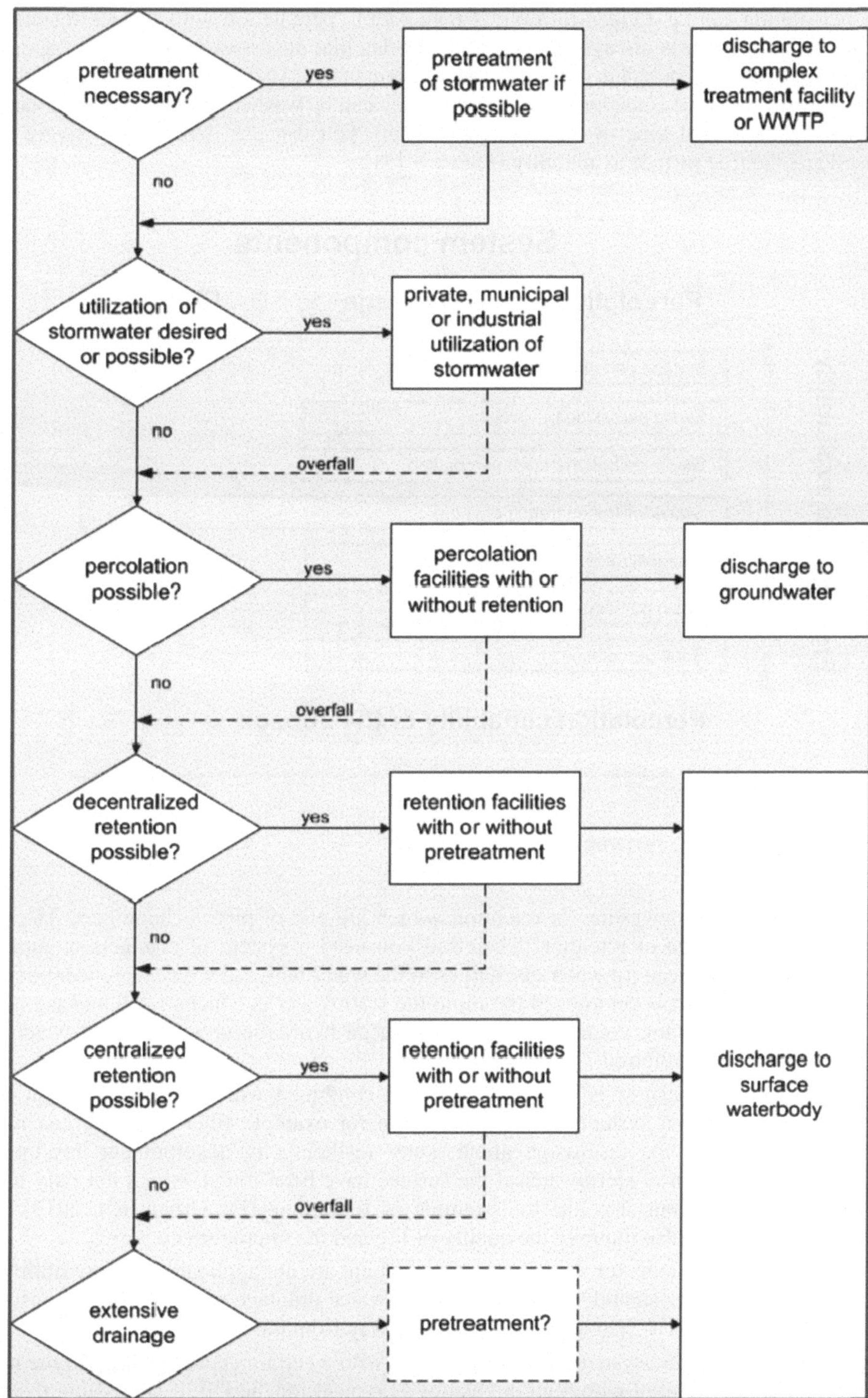

Figure 8.12 Elements of stormwater management: a decision support diagram.

The utilization of stromwater for private, municipal or industrial purposes is a very important element of stormwater management. Especially big areas like roofs of stadiums, industrial complexes etc., can collect a significant amount of rainwater, which can for example be used for toilet flushing, cleaning or as process water. Thereby the runoff is significantly reduced. A lot of ideas for the utilization of stormwater have been developed through the recent years, and are often published in the context of rainwater harvesting. Such ideas should be supported and implemented because they can reduce the demand for water of drinking water quality significantly.

The measures for percolation can be divided into direct percolation, percolation with surface retention and percolation with subsurface retention. Stormwater is always loaded with particles like dust, soil or harmful substances. Depending on the load of the rainwater it can be percolated if the soil is permeable. Also, the load of harmful substances which accumulates in the soil always has to be considered and whether they can be washed out to the groundwater or if they can be deposited by the wind. Which design type of a percolation system you can use depends decisively on the percolation capability of the subsoil and the free surface availability (Figure 8.13).

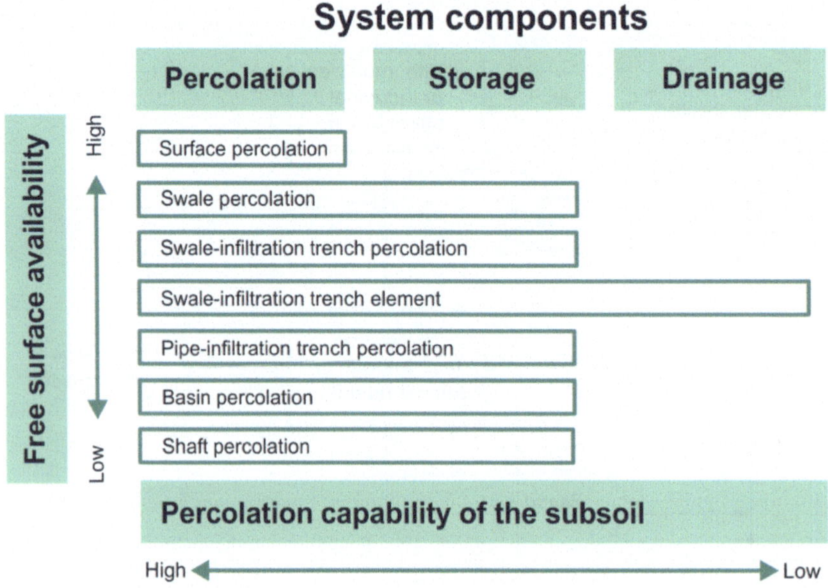

Figure 8.13 Design types of percolation systems after DWA (2005).

Percolation is often combined with measures of retention, which are end of pipe technologies. The stormwater always has to be collected and led to the area of retention. Therefore you need a system of channels or pipes which has to be maintained. From the retention area, the stormwater discharges to the water body, so we get a concentration of the discharge at a certain point. The more the system is centralized the more the stormwater is concentrated and the higher the impact to the water body or the necessary retention volume to secure a reduced hydraulic pressure to the water body. Therefore a more decentralized system should be preferred.

The advantage of all retention measures is they can easily be combined with a (pre-) treatment of the stromwater. Possibilities for the design of retention systems in settlements are for example filter basins, green roofs or stormwater storage tanks. During the last years the discussion about water resilient city development has intensified. New and innovative approaches for the retention of stormwater at the surface have been found, which not only include the usage of the storage volume of parking areas but also like the example of Rotterdam (De Urbanisten, 2013) demonstrates, well designed open spaces in the cities can also improve the quality of life and the microclimate.

Only if all the other mentioned elements for stormwater management are not applicable, the possibility of building up an excessive stormwater drainage system should be checked. Stormwater drainage systems are expensive when built and during maintenance and the stormwater is channeled to a place far away from the place the rain falls down.

All the above mentioned elements and systems can only be built with a certain capacity and also the maximum inflow of water is limited. During periods of rainfall with high intensities as typical for the DF, it is possible that stormwater leaves the system and provokes a flooding event at the adjacent areas. Otherwise it is neither technically nor economically reasonable to build systems for very extreme events. Nonetheless, it is necessary to discuss, especially under the aspect of climate change, which stormwater events should be used for the calculation of the systems and to determine when it is allowed that the systems be overstrained. It is one of the most important facts during the planning of a system to think about what to do with the overfall. There are possibilities like leading it to an open area or to a sewer system, but the best solution depends on the local conditions and the decision has to be found on a by-case basis.

Another very important fact is to plan and execute the necessary maintenance work during the different times of the year. This is the only way to secure the function and reduce the risk of damages by water logging or flooding. This is especially important because the different elements of the stormwater management are not operated by one institution. The utilization of stormwater is facing the challenge of a high number of households, that is, private persons, which are the operators of systems for percolation or decentralized retention. Since organizations responsible for the drainage system will have a big interest that the different systems will and therefore, it will be necessary to train people operating the systems by giving them information and stipulating a specification sheet.

8.3.2 Stormwater modeling at a small catchment of Brasília

The correct prediction of flooding in urban areas is an important challenge to fulfill public regulations. The sewer network is one of the key factors to assure a safe drainage of urban areas and to avoid claims caused by urban flooding. But when discussing the impact of urbanization and climate change a sole consideration of the sewer network is not sufficient, the impact of the overfall of the system described in Chapter 8.3.1 and their impact on urban runoff has to be considered. Therefore, sewer simulations are an option. But traditional sewer simulations deliver only the basic information for a rudimental flood protection and almost no information about what happens on the surface with the water before it enters the sewer system.

To get more information it is necessary to consider the interaction between sewer and surface runoff, which is only possible by a bi-directional modelling, as already described by Djordjevic *et al.* (1999) and Schmitt *et al.* (2005). This kind of modeling regards the infiltration in the sewer from the surface and the exfiltration of the sewer to the surface in every calculation time step (Figure 8.14). This continuous exchange between the two flow regimes (surface and sewer) makes the verification of the flood protection of urban areas possible. The two regimes are connected via inlets and grates, which only allow a certain exchange rate.

Therefore, detailed information about the relevant structures on the surface is necessary. One very important point to consider is the quality of the surface-model which is used for the calculation of the surface flow. A lot of detailed physical information like height of curbs or site of street inlets is necessary. Laser scan data is expected to come close to the necessary detail of information and could serve as basis for modelling of digital terrain models (DTM). But even this data has variances which does not allow an unmodified use especially when small structures like curbs are of importance and a high reliability is asked. Therefore the data has to be verified and detailed by field inspections with further levelling. The complex context of urban drainage and the DTM make it necessary to cope with a plenitude of information and data which leads to high requirements on computer capacity and performance of the models used.

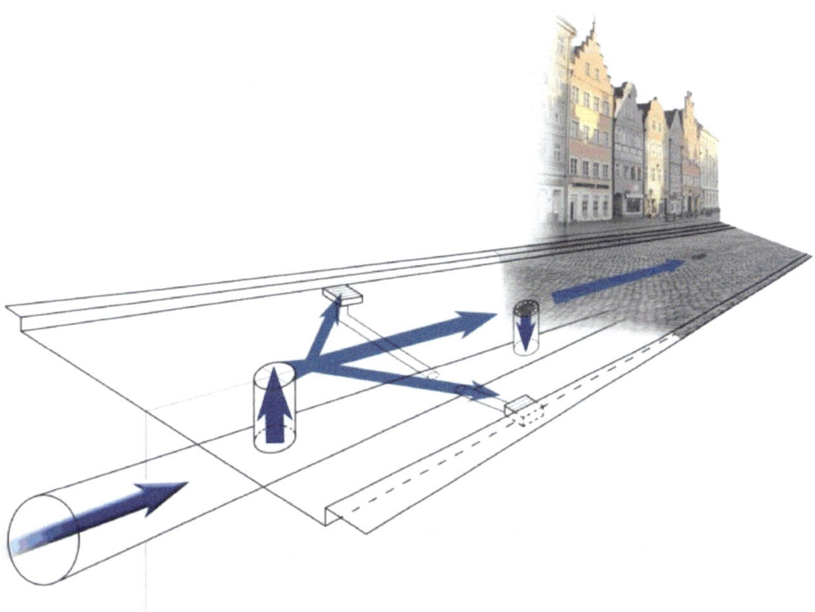

Figure 8.14 Bi-directional calculation of sewer and surface.

After all, this kind of model can only give information about what happens after the water has entered and left the sewer system. But this might be not beneficial because it is also information needed about the water on the surface during the runoff process, because flooding can already occur when the water does not enter the sewer system. This happens especially during extreme stormwater events. Standard sewer simulations do not have this option because ususally the inflow is simulated hydrological. Therefore, if a simulation of flooding caused by heavy rainfall is necessary, the simulation of inflow to the sewer hydraulic by using the St.Venant equations with a detailed 2D-model of the surface is necessary.

The 2D-model of the surface consists of a plentitude of triangles which are connected via their sides. An exchange of water is possible form one triangle to the other via these sides. Also every triangle can get water from an external source. It is possible to define a time variation curve for this external source, and if the time variation curve is identical with the rainfall, it is possible to calculate the runoff process of the water on the surface. Therefore, several modifications are necessary, and an excessive calibration of the model should be carried out, which makes measurements of good quality in the sewer system necessary. A great number of data is required for this kind of modelling. For a qualified standard sewer calculation the following informations are needed,

- pervious and impervious areas (location!)
- land use
- soil (infiltration!)
- stormwater drainage system (material, location, age, …)
- rainfall-data.

For a more detailed advanced sewer calculation like described above additional information are needed:

- exact and dense data of the surface, for example, LiDAR
- location and height of curbs
- location and kind of inlets
- for flood affected areas aggregation of (surface-)data
- location and dimension of vertical structures (walls etc.)
- location and dimension of flood relevant structures, for example, basement garages.

Figure 8.15 Catchment area in Brasília for advanced sewer simulation.

For a small catchment (Figure 8.15) of Brasília it was planned to do this advanced sewer calculation including a 2D calculation of the surface runoff. For the simulation the program ++Systems was used. The modeling with this tool is described in Obermayer *et al.* (2010). The catchment is 36 ha (800 m × 450 m) and is surrounded by tall multistoried buildings. Typical for Brasília are the multilane roads which enclose the catchment. In the catchment a lot of parking areas are located which are almost completely sealed. The catchment has a base slope of 3.6 percent. Most of the buildings are divided by a green corridor with intensive vegetation, which could be used for percolation facilities (see Figure 8.16). There is no surface inflow to the catchment from outside. The stormwater is now collected in a branched stormwater drainage system with diameters from 300 mm to 120 mm and is collected with curb-opening inlets which are integrated in the 15 cm high curbs.

Figure 8.16 Combination of sealed road surfaces and green areas of Brasília.

Like shown above for such kind of modeling, a lot of data is required. Unfortunately the information about the surface of the selected catchment is insufficient to finish the calculation. Also, the data for a calibration of the model is missing. However, calculations in another catchments in Germany have shown that with such kind of modeling it is possible to create information needed to locate endangered areas and to develop strategies to protect them. It is also possible to consider the behavior of percolation systems or retention areas by heavy rainfall events. The calculation for such a small catchment lasts about 12 h (with a standard Quadcore PC). This calculation times make it possible to calculate different stormwater events with different intensities and time variations. Thereby it is possible to make a multi-case analysis for the vulnerability of the affected areas and find optimal solutions for the complex interaction of sewer systems, surface and all other elements of urban drainage to create a sustainable flooding protection.

8.4 CONCLUSIONS AND RECOMMENDATIONS

The example of Brasília shows that an examination of topics such as wastewater treatment and stormwater management are absolutely necessary with respect to future challenges such as demographic and climate changes. The results of the pilot plant for advanced wastewater treatment are supporting the idea of removal of hazardous microsubstances. The next step towards a use of the lake as a drinking water source would be further research on the topic of advanced treatment techniques for Brasília and a large-scale implementation. Additionally, it is also important to identify incorrect and illegal connections and damaged areas of the sewer and drainage system, to prevent entry of polluted and non-treated water into the water body and hydraulic overloads of the existing treatment plants, which might lead to low qualities of discharge.

The stormwater management in Brasília is well developed, but flooding events have shown it is important to reconsider the existing system. It is a requirement to get more information about the quality of the stormwater runoff durin the course of the year. The rainfall variability makes it likely that after the longer dry periods there is a runoff with a high load of potentially harmful substances especially from the very busy multilane roads. To decide what measurements can be taken to improve the stormwater management it seems advisable to categorize the quality of the runoff for different areas of Brasília.

In the future the existing drainage system should be regarded only as a 'very important' part of the concept for stormwater management. The existing drainage system can be supplemented by a variety of systems for utilization of stormwater, percolation or decentralized or centralized retention. Especially an increase of percolation in the whole city seems advisable. Plants for the management of stormwater can be integrated as a visible part of the city development. Therefore, contact with the responsible authority should be intensified. Flood protection is a big challenge for the future. To find optimal solutions for the complex interaction of sewer systems, surface and all other elements of urban drainage advanced simulations of the affected areas may be necessary.

8.5 REFERENCES

Abegglen C. (2009). Spurenstoffe eliminieren: Kläranlagentechnik. Eawag News 678d/Juni 2009. S. 25–27.

Abegglen C. and Siegrist H. (2012). Mikroverunreinigungen aus kommunalem Abwasser. Verfahren zur weitergehenden Elimination auf Kläranlagen. Bundesamt für Umwelt, Bern, Umwelt-Wissen Nr. 1214:210.

Aster S., Vasyukova E., Ripl K., Kuse B., Günther N., Uhl W., Makeschin F., Wummel J., Krebs P. and Günthert F. W. (2010). Developing an Integrated Water Resources Management (IWRM) Concept for the Capital Brasília. gwf-Wasser Abwasser. International Issue 2010, 52–57.

Bester K. (2008). Personal Care Products in the Environment. Wiley-VCH, Weinheim.

BLAC (2003). Bund/Länderausschuss für Chemikaliensicherheit (BLAC). Arzneimittel in der Umwelt. Auswertung der Untersuchungsergebnisse. Bericht an die 61. Umweltministerkonferenz am 19./20. November 2003. Hamburg.

Cornel P. and Meda A. (2009). Weitergehende Behandlung von Kläranlagenabläufen. Wasser-Wirtschafts-Kurse Kommunale Abwasserbehandlung, 7.–9.10.2009.

Cornel P. (2010). Eignung von Aktivkohle zur Abwasserbehandlung. Aktivkohle in der Abwasserreinigung, vom Versuch zum technischen Maßstab. Tagungsband Symposium Aktivkohle, DWA, Stuttgart.

DE Urbanisten (2013). Water cycle based urban plans to flood defense. http://www.urbanisten.nl/wp/ (accessed 10 July 2013).

DWA (2008). Anthropogene Spurenstoffe im Wasserkreislauf – Arzneistoffe. DWA-Themen der Deutschen Vereinigung für Wasserwirtschaft, Abwasser und Abfall e.V., Hennef.

DWA (2006). Leitlinien der integralen Siedlungsentwässerung (ISiE) – DWA Regelwerk A100. Deutschen Vereinigung für Wasserwirtschaft, Abwasser und Abfall e.V., Hennef.

DWA (2005). Planung, Bau und Betrieb von Anlagen zur Versickerung von Niederschlagswasser- DWA Regelwerk A 138. Deutschen Vereinigung für Wasserwirtschaft, Abwasser und Abfall e.V., Hennef.

Djordjevic S., Prodanovic D. and Maksimovic C. (1999). An approach to simulation of dual drainage. *Wat. Sci. Tech.*, **39**(9), 95–103.

Fahlenkamp H. *et al.* (2006). Abschlussbericht zum Forschungsvorhaben 'Untersuchungen zum Eintrag und zur Elimination von gefährlichen Stoffen in kommunalen Kläranlagen, Teil 2'. Ministerium für Umwelt und Naturschutz, Landwirtschaft und Verbraucherschutz des Landes Nordrhein-Westfalen.

Friedrich H. (2005). Untersuchungen zum Eintrag und zur Elimination von gefährlichen Stoffen in kommunalen Kläranlagen. In: Membrantechnik in der Wasseraufbereitung und Abwasserbehandlung – Perspektiven, Neuentwicklungen und Betriebserfahrungen im In- und Ausland. Begleitbuch zur 6. Aachener Tagung Siedlungswasserwirtschaft und Verfahrenstechnik.

Gujer W. (2008). Siedlungswasserwirtschaft.Springer-Verlag Berlin Heidelberg.

Ivashechkin P. (2006). Elimination organischer Spurenstoffe aus kommunalen Abwasser – Dissertation. Institut für Siedlungswasserwirtschaft, Bd. 205, Aachen.

Joss A., Siegrist H. and Ternes T. A. (2008). Are we about to upgrade wastewater treatment for removing organic micropollutants? *Water Res.*, **40**(8), 1686–1696.

Kloepfer A. *et al.* (2005). Occurrence, sources and fate of benzothiazoles in municipal wastewater treatment plants. *Environ. Sc. Technol.* **39**(10), 3792–3798.

Knepper T.P. *et al.* (2004). Removal of persistent polar pollutants through improved treatment of wastewater effluents (P-THREE). *Water Sc. Technol.*, **39**(10), 3792–3798.

LANUV NRW (2007). Eintrag von Arzneimitteln und deren Verhalten und Verbleib in der Umwelt – Literaturstudie. Studie im Auftrag des Landesamts für Natur, Umwelt und Verbraucherschutz des Landes Nordrhein Westfalen, Recklinghausen.

Obermayer A., Günthert F. W., Angermair G., Tandler R., Braunschmidt S. and Milojevic N. (2010). Different approaches fort he modelling of sewer caused urban flooding; *Wat. Sci. Tech.*, **62**(2), 175–182.

Reemtsma T. and Jekel M. (2006). Organic Pollutants in the Water Cycle – Properties, Occurrence, Analysis and Environmental Relevance of Polar Compounds. Wiley-VCH Verlag, Weinheim.

Schmitt T. G., Thomas M. and Ettrich N. (2005). Assessment of urban flooding by dual drainage simulation model RisUrSim. *Wat. Sci. & Tech.*, **52**(5), 257–264.

Schrader C. (2007). Verfahrenstechnische Möglichkeiten zur Entfernung von organischen Spurenstoffen aus kommunalem Abwasser – Dissertation. Institut für Siedlungswasserbau, Wassergüte- und Abfallwirtschaft, 190, Stuttgart.

Ternes T. and Joss A. (2006). Human Pharmaceuticals, Hormones and Fragrances – The challenge of micropollutants in urban water management. IWA Publishing, London, New York.

Chapter 9

Capacity development as base element of Integrated Water Resource Management in Central Brazil

B. Kuse, J. Wummel and K. D. Neder

9.1 INTRODUCTION

In frame of the IWAS-ÀGUA DF project, Sachsen Wasser GmbH (SaWa), Kommunale Wasserwerke Leipzig (KWL – water and sanitation company of Leipzig, SaWa's parent company) and Companhia de Saneamento Ambiental do Distrito Federal (CAESB – water and sanitation company of the Brasília DF), have signed a Technical Cooperation Agreement with the aim to promote capacity development at CAESB. The United Nations Development Programme (UNDP, 2008) defines three levels of capacity development.

- The overall level, which determines the 'rules of the game' for the interaction between and among organizations through policies, legislation, power relations and social norms,
- The organisational level, which comprises the internal policies, arrangements, procedures and frameworks that allow a utility to operate and deliver on its mandate, and that enables the coming together of individual capacities to work together and achieve goals, and
- The individual level, which refers to the skills, experience and knowledge of people within the utility.

In consequence, it was assumed each actor is endowed with a mix of capacities that allows to perform them. Some of these are acquired through formal training and education, others through learning by doing and experience. The main objective of this cooperation was to develop practically orientated elements, for example, methods to find an investment strategy or implementation of a remote control system for the integrated water resources management (IWRM) for the Distrito Federal. Apart from technical solutions, also improvement of management methods and decision support systems were discussed. Because SaWa and CAESB are municipal water supply and sanitation utilities, the focus of this cooperation was set on the transfer of management experiences and capacity development, both on organisational and individual levels.

Within the framework of the project, the transfer of management expertise was achieved through intensive dialog-orientated discussions. In this sense, KWL's experiences, especially those derived during the time after German reunification, seemed to be tailor made for CAESB, as CAESB was facing similar transformation processes.

Therefore, during the planning of activities, emphasis was put on the fact that participants of both companies were experts possessing extensive experience in the implementation of water supply and sanitation processes. These experts identified several topics, where CAESB planned to implement new processes (e.g. data analysing) and KWL had already gained wide experience. The partners, thus, agreed that these topics should be the main focus for the capacity development for the DF.

Capacity development that includes individuals and organisations can produce positive changes in the water and sanitation sector. Effective capacity development can ensure a continuous supply of water of high quality, which is being continuously offered to customers, and also provide a proper wastewater treatment at reasonable costs. However, capacity development and especially human capacity development is not a 'quick fix'. Capacity development needs a culture within a company, where open discussion is desired and mistakes are permitted. However, it may take sometimes years, to achieve full implementation. Therefore, the most important element of capacity development is the commitment of the involved participants and of the organisational leader, including the political will. Throughout the cooperation, although there was a change in the position of the president of CAESB, there was no change of internal policies on capacity development.

9.2 SHORT PRESENTATION OF THE PARTNERS IN WORKING GROUP CAPACITY DEVELOPMENT

9.2.1 Companhia de Saneamento Ambiental do Distrito Federal (CAESB)

CAESB is based in Brasília, Brazil, and is a public owned company, which belongs to the Government of the DF of Brazil. CAESB has been recognized as one of the best water and wastewater utilities in Brazil. The company serves a total of more than 2,606,000 people in the region of the DF, covering all the existing urban and rural agglomerations and communities. The experience in water and wastewater management in the Brasília DF began with the construction of Brasília city since the last 50 years. At present, the company supplies around 234 million m of potable water every year, while approximately 110 million m of wastewater are treated. Annual turnover from the water and wastewater services is about 360 million EURO.

Today, CAESB operates a total of 9 water and 16 wastewater treatment plants. The largest wastewater treatment plant has a treatment capacity of around 129,600 m per day. The total length of the water distribution system amounts to ca. 8377 km, including pipe diameters from DN 80 to DN 2400 and 40 water pumping stations. The length of the sewerage system equals ca. 5112 km; it includes 54 wastewater pumping stations. CAESB is responsible for all operation, maintenance and repair services related to the water supply utilities and networks, as well as those related to the wastewater systems. Water loss has declined from roughly 30% in 2007 to approximately 24% (2012) due to the implementation of a water losses reduction program. The current number of drinking water connections served is over 580,000, with 99.8% of the customers metered. The collection rate for waste water exceeds 95%.

CAESB currently employs ca. 2300 qualified professionals to fulfil all obligations from the management contract that exists between the DF and the company. The management contract comprises all aspects of the technical and commercial management of the water and wastewater system including metering, billing, revenue collection and investment planning. CAESB aims at being a modern, market-based and customer-oriented company having cost-covering prices and modern management tools. CAESB has been acting in commercial, financial and technical consulting within the water supply and sanitation sector in Brazil and abroad. The company transfers its know-how to partners and clients dealing especially with low cost sanitation solutions for developing countries.

9.2.2 Sachsen Wasser (SaWa) and Kommunale Wasserwerke Leipzig (KWL)

SaWa is a 100% subsidiary company of KWL, which is one of the largest water and wastewater utilities in Germany. Both companies are based in Leipzig, Germany. KWL is a public utility owned to 100% by the city of Leipzig and its surrounding 15 municipalities.

KWL serves a total of more than 628,000 people in the region of Leipzig, covering the urban agglomeration and the surrounding communities. In addition SaWa supplies water to 45,000 people near Berlin. The experience in water and wastewater management in the Leipzig area dates back to over 100 years. KWL supplies around 32 million m of potable water every year, while approximately 45 million m of wastewater are treated. The annual turnover from the water and wastewater services was about 133 million € in 2012.

Today, KWL operates a total of 5 water works and 22 wastewater treatment plants; the largest WWTP has a treatment capacity of around 120,000 m per day. The length of the water distribution amounts to ca. 3300 km, including pipe diameters from DN 80 to DN 1200 and pumping stations. The length of the sewerage system equals ca. 2750 km and includes 205 wastewater and storm water pumping stations, as well as 134 storm water treatment (settlement) facilities. KWL is responsible for all operation, maintenance and repair services related to the water supply utilities and networks, as well as those related to the wastewater disposal systems. Water losses have declined from ca. 25% in 1995 to approximately 12% in 2001 due to the increased level of expertise in leak detection as well as operation and maintenance of water mains.

The number of drinking water connections served reaches 80,000 which include 4500 non-domestic customers. Almost 100% of the customers are metered. The wastewater collection rate well exceeds 90%. Due to this success, the KWL Group has been able to extend its revenue collection services to other municipal companies in the Leipzig area, notably the public transport utility.

KWL currently employs ca. 540 highly skilled professionals to fulfil all obligations according to the management contract that exists between the municipality of Leipzig and the surrounding municipalities and the company. The management contract comprises all aspects of the technical and commercial management of the water and wastewater system including metering, billing, revenue collection and investment planning. SaWa has access to all technical and commercial expertise gathered within the KWL group. Whereas SaWa itself has 36 staff members, the total staffing of the KWL Group amounts to approximately 800. Since 1990, KWL has successfully undergone transformation from a formally centrally structured water distribution agency in the socialist former German Democratic Republic (with hierarchical structures, highly subsidized tariffs and high rates of non-revenue water) to a modern, market-based and customer-oriented company with cost-covering prices and modern management tools. Therefore, the company possesses a large experience in the transformation of a public water and wastewater entity to a market-based company including all aspects of both technical improvements and the introduction of customer service and business management tools.

SaWa is specialised in commercial, financial and technical consulting within the water and sanitation sector. The company transfers its know-how on modern operating procedures for water supply and waste water disposal to partners and clients. With this, it supports municipalities to be more effective and efficient.

It supports and applies solutions in which municipalities assume responsibility for their water management. For many municipalities as well as water and wastewater associations the municipal-based scheme implemented in Leipzig represents an efficient alternative to the pure water management privatisation. This was one of the reasons why CAESB expressed a high interest in working with SaWa.

9.3 APPROACH

Improving and performing capacity development was the main goal of the cooperation between SaWa und CAESB, and remained the main focus even after the position of the president of CAESB has changed. Furthermore, this approach supported the other (German) project partners within the IWAS-ÁGUA DF project to find, within CAESB, open and interested experts who were willing to address changes and innovations.

In the assessment of CAESB, its deficits may mostly be found in personnel qualification. CAESB has a great interest in the so-called innovative management methods. This was the starting point for the selection of topics for the different workshops of the first project phase (January 2010 to December 2011). Planning common activities emphasized that participants from both partner sides were experts in the implementation of major processes of water supply and wastewater disposal.

During the first phase of the cooperation, three one-week workshops (2010–2011) focussing on various topics, such as drinking water supply, wastewater disposal, and management processes were conducted in Leipzig, Germany. During these workshops, experts from KWL/SaWa demonstrated their working methods and experiences in the major processes of water supply and wastewater disposal (see chapter 9.4). The workshops consisted of a mix of presentations and discussions. Similarities and differences in the approach in handling of CAESB and KWL were identified. Additionally, the practical part of the cooperation was enlarged by extensive site visits to water treatment facilities.

Every workshop in the first project phase was characterised by a variety of single topics to build a solid basis for professional discussions. At the same time current projects of CAESB were explained to SaWa. In some of them KWL had made extensive experiences during implementation. So the partners agreed to focus the common activities on specific topics in the second phase (April 2012 to June 2013). They decided to continue the exchange of experiences mainly through two workshops regarding these thematic priorities (see chapter 9.5). As a result the transfer of experiences was significantly intensified and the transferability of these experiences to the conditions of CAESB was secured.

9.4 SHARING THE EXPERIENCE IN WORKSHOPS (2010–11)

The everyday life of a water supplier is characterized by an abundance of single activities. The specific approach in the company arises from a number of framework conditions, which in turn are defined by previous development and the current requirements and capacities of the company. The base for transferring experiences between both companies is the presentation and discussion of specific problems for a given situation and the experiences for solution of these problems. It became clear to the experts that the comparison of each corporate development is a prerequisite for a purposeful exchange of experiences. Based on the first phase of the subproject 'Capacity Development' in the period from January 2010 until December 2011. A structure was determined, that successively covered the company departments for drinking water supply, wastewater disposal and auxiliary processes of administration. For each of these departments a one-week workshop in Leipzig was organized. Participants of these workshops were CAESB experts who hold responsibilities in the middle-level management for processes listed above. Programs were adjusted to the participants by selecting specific presentations, professional discussions and site visits to facilities.

During the workshop regarding the 'wastewater department', KWL and SaWa experts presented the following topics using presentations with intense professional discussions.

- Basics of investment decision at KWL on the example of a wastewater treatment plant situated near Leipzig
- Operational processes in the sewer network of KWL
- Introduction of a process control system in the wastewater system of KWL in the 1990s.

The related professional discussions were intensified during site visits to different wastewater treatment plants, to the central control room, to the wastewater laboratory of KWL and the demonstration of channel inspection and cleaning.

During the workshop regarding the 'drinking water supply department' the presentations, with intense professional discussions, comprised following topics.

- Basics of an investment decision at KWL on the example of a waterworks of KWL
- Operational processes in the drinking water supply system of KWL
- Introduction of a process control system in the drinking water supply system of KWL in the 90s.

Related discussions were continued during site visits to several waterworks, to the central control room and to the water quality control laboratory of KWL.

For the following workshop several topics were offered by SaWa. CAESB selected those topics, which were related to topics in administration. Again, these topics were presented to the CAESB experts by experts of KWL and SaWa using presentations with intense professional discussions. The following topics were included.

- Introduction and use of different IT applications at KWL (including failure management)
- Experiences in tariff calculation and different tariff systems at KWL and CAESB
- Customer Service at KWL, example of new connections and the use of prepaid water meters
- Experiences at KWL respectively SaWa regarding the topics 'center calculation' and the application of personal indicators for performance-related salary components

The final workshop took place in Brasilia, where both KWL and SaWa staff noticed that several suggestions from the former workshops were directly adopted in current activities of CAESB. This was the reason why CAESB asked KWL to continue the exchange with focus on current CAESB's projects.

9.5 SPECIFIC TOPICS (2012–13)

9.5.1 Background of the second project phase

During evaluation sessions in Brasilia based on the results of the first phase of the subproject new fields of activity, mostly related to current projects at CAESB on the implementation of new processes, were identified. For some of these fields KWL possessed extensive experience. The implementation of these new topics will give a high potential of support for the total project IWAS-ÁGUA DF. Topics that are described hereafter were then implemented depending on particular framework conditions and requirements, taking care that transferability of these experiences to the conditions of CAESB were ensured.

9.5.2 Implementation of a laboratory information management system

The presentation of the IT software landscape of KWL illustrated how analytical results from both central laboratories of KWL (drinking water and wastewater) are not only automatically collected in a laboratory information management system (LIMS) but also are entered in the respective report systems. This automated process of assignment from analysis results to the corresponding sample attracted a big interest among the CAESB's laboratory staff, because the company is also working on the implementation of a LIMS. Due to the advanced processing stage at CAESB a less time-consuming form of cooperation was selected. A detailed report on the approach of KWL regarding the administration of analytical data was prepared for CAESB. Further specific inquiries could be reasoned out by direct transfer of contacts between the respective staff of KWL and CAESB.

9.5.3 Implementation of a data management system

In all workshops of the first project phase the automation system of KWL including central process monitoring, administration, allocation of dates, databases, and LIMS played a central role. During the introduction of the respective system that took several years, KWL gained extensive experiences. Therefore, CAESB initiated an exchange of more detailed information concerning data management in the second project phase. Since several years, CAESB has been setting up a central process monitoring for its drinking water facilities and wastewater facilities. Therefore, by starting such a discussion CAESB could benefit directly from KWL's experiences and improve the implementation of centralized process monitoring.

In a first step, experts for a professional discussion between KWL and CAESB were selected. Participants of a one-week workshop on 'Data Management' were selected experts in 'technology', 'automation technology' and 'reporting'. The objectives and expectations regarding this workshop as well as the specific program were determined in advance together with the participants. Through presentations with intense discussions following topics were considered by the KWL experts in Leipzig.

- Implementation of automation ('from a measured value to the process mapping')
- Use of process control system and data archiving
- Preparation of reports with different objectives (accordingly as seen from the perspective of the operational staff, the technician and the leader of the central control room)

The professional discussions were intensified during site visits to different wastewater treatment plants and the central control room. The huge number of detailed questions from CAESB experts showed the intention of CAESB to apply the experiences made in Leipzig.

9.5.4 Energy efficiency measures

The third topical focus within the subproject 'Capacity Development' was the implementation of methods for energy saving measures for CAESB facilities. SaWa and KWL realised a number of energy analyses in this area and took part in different benchmark events. At the same time, CAESB promotes activities in the field of (electric) energy efficiency.

During a one-week professional exchange in Brasília, the previously implemented projects of SaWa and KWL to raise energy efficiency and the current activities to introduce an energy management system according to international guidelines were outlined. For application, an energy analysis for the CAESB wastewater treatment plant 'Brasília North' was implemented. From a methodological point of view, the data collection and the reliability assessment were in focus of the common activity within this project. Another focus was the evaluation of the installed technology and aggregates regarding energy efficiency and security connected to wastewater treatment.

In connection with the preceding professional exchange in Leipzig, CAESB was able to compile an own concept for installation of the required performance measurements. In the future this will allow CAESB a more reliable energy efficiency assessment at their main facilities and facility parts.

9.6 RESULTS AND CONCLUSIONS

The cooperation with CAESB proved to be very advantageous for the project, especially due to the fact that a department for special projects (*Assessoria de Projetos Especiais, PRE*) already existed at CAESB. This department is well integrated into the corporation and has positive experience in project implementation. Therefore, there was a properly integrated project partner who could identify the responsible expert for each discussed topic. During the project implementation it turned out that CAESB, apart from repetitive tasks of a water and sanitation service provider, has good experiences in project implementation methods. Despite numerous differences between both municipal water companies (KWL and CAESB, see chapter 9.2) several common topic areas could be identified. In this way it was possible to provide CAESB with experiences, which KWL gained itself during its reorganization to a modern water supply and sanitation company.

According to the goals of the subproject "Capacity Development", the capability of the Brazilian project partner CAESB to be an efficient water supply and sewerage company was strengthened. The benefit for the total process is well illustrated by the implementation of the lessons learned of the second project phase (see chapter 9.5.). The central laboratory of CAESB is well equipped with analytical equipment. However, the administration of analytical data needs still improvement. Thus, the introduction of a laboratory information system is imminent. The setup of a central remote control for the drinking water supply and wastewater treatment facilities is usually a protracted task in bigger companies. KWL integrates all locations in the central remote control room. All the locations can be supervised and, if needed, remotely operated. Furthermore, with the data archiving an instrument was created that provides process data. For this purpose, process data together with analytical data and further operational data are stored in a common database. The latter is organised as an open system to enable a simplified preparation of reports by all involved departments. CAESB intends to follow this way but is still in an early phase. The company depicted the crucial functions of all main facilities in the central control room and supervises the facilities from there. The provision of data processing in all interested company departments is an essential precondition for a modern supply company. At this point the cooperation during the project work package 'Capacity Development' started and the necessary experiences where conveyed.

In addition to the results of the project IWAS-ÁGUA DF, CAESB and SaWa will continue to cooperate. The idea is to use the specific knowledge of each partner to problems, which still needs to be addressed. Moreover, SaWa starts communication with IFG (Instituto Federal de Goiás, Goiania, Brazil) to find a way, to support local companies to increase sustainability of their activities. So far, financing of these activities is the greatest challenge.

9.7 REFERENCE

UNDP (United Nations Development Programme) (2008). Practise Note: Capacity Development.

Chapter 10

Lessons learned – conclusions

C. Lorz, F. Frimmel, W. Günthert, S. Koide, F. Makeschin, K. Neder, W. Uhl,
D. Walde, H. Weiss and E. Worch

10.1 PROCESSES IN RIVER BASINS

Climate is a major factor influencing availability and quality of water resources in the study region. The specific semi-humid conditions, that is, uneven annual distribution of rainfall during dry and wet season, and its spatio-temporal dynamics, are of particular importance for the IWRM approach. For the analysis of climate conditions the database CLIMA-DF was developed and input data were tested for errors and homogeneity. The database was used to obtain the baseline climate as reference. For the period 1971–2000 a considerable spatial variability of temperature and rainfall was found reflecting the topography of the region. For the period 2001–2010 a clear trend towards higher mean annual temperatures and a longer dry season was found. The analysis of General Climate Models, aggregated to regional scenarios by using statistical downscaling, showed a continuation of this trend for the future climate of the Distrito Federal. Although, there are still some uncertainties for predicting the future water availability, there is a clear trend towards less available water during the in future longer dry season.

    Groundwater so far, has rather not been taken into consideration in the Distrito Federal, that is, only 4% of the direct total water supply comes from groundwater. However, because the majority of streamwater during dry season is baseflow from groundwater and increasing water demand due to population growth and urbanization, groundwater is gaining importance for water supply. Therefore, the development of best management practices for groundwater resources is needed. However, it requires fundamental knowledge about aquifers, for example, structural setting, storage capacity, recharge mechanisms and groundwater flow dynamics. Exemplarily, a numeric model was developed for the meso-scale river basin Pipiripau based on existing concepts of the hydrogeological system. The approach might be used as base for groundwater protection and management concepts such as the delineation of groundwater protection zones, vulnerability mapping, groundwater extraction, land-use optimization or future climate driven scenarios. In a small-scale study an approach on the assessment of pollution potential from a municipal landfill adjacent to the watershed of a water reservoir was developed. Geoelectrical tomography and Direct-Push sounding and sampling was used to identify, localize and assess the potential for groundwater contamination. An immediate threat of the groundwater in the watershed could be excluded. A third case study aimed at the groundwater in the wider region of the town São Sebastião, where water supply is entirely based on groundwater. Stable isotope mapping was used to identify the driving recharge mechanisms of the study region. In addition, data obtained by monitoring were analyzed to test for possible overexploitation of the local aquifers. Finally, an additional small-scale study dealt with the idea of artificial groundwater recharge by using pre-treated wastewater for infiltration. An approach was developed to increase recharge rates during dry periods, droughts or in case of overexploitation. From all case studies it can be concluded that groundwater is a (potential) source for drinking water supply in the region. For its sustainable use a concept of groundwater management must be developed including monitoring networks, modelling approach and sustainable exploitation concepts.

    Water resources and **land use** is a major complex with high importance for the study region because water supply depends largely on collection of surface water and changes in land use have been rather dramatic during the last decades. Land use effects caused in the meso-scale river basin Pipiripau, part of the national pilot project on environmental services (*produtor de água*), and other river basins a dramatic decrease of baseflow discharge during dry season, when water demand is highest, mostly due to irrigation. For the Pipiripau river basin a meso-scale simulation tested the efficiency of small basins and terraces for sediment retention and linked this to costs. The outcome showed the optimal ratio of investments for measures and benefit in terms of reduction of sediment generation. For the same river basin the web-based planning support tool *Letsmap do Brasil* was developed aiming at the visualization of land use change and its effects, transparency and user friendliness. The system will enable stakeholders with different backgrounds to participate in planning and management processes. Finally, a geo-footprint approach was developed to identify major sediment sources in the basin of the Paranoá Lake. Bare soils and construction sites in urban areas were identified as major source areas. Thus, major challenges in reducing/preventing silting of reservoirs will be the rehabilitation of bare (degraded) soils and a legal framework to reduce sediment generation especially from construction sites. In general, for the land-use-water-complex as a crucial element of the IWRM approach we were able to show exemplarily the integration of data from regular monitoring as well as field data, lab data and simulated data using numeric models and planning support tools.

Finally, **urbanization** is the major driver for changing frame conditions of water resources in the study region. The prediction of water demand and effects of urbanization on water resources is a major challenge for the future, which will be solved by using modern approaches integrating social, technical and scientific knowledge only. For three characteristic regions around Brasília an Urban Structure Type approach was developed. The approach merged socio-economic information from census data, remote sensing data and water related information. As major outcome, a map of urban structure types was produced and might be used as a future base for assessing urban areas in scope of IWRM.

10.2 WATER QUALITY IN WATER RESERVOIRS

For the first time a comprehensive water quality assessment with specific focus on organic micropollutants was carried out for Lake Paranoá. With consideration of existing long-term data as well as information on urban structures of the lake basin, the obtained results allow to attest a good to very good hydrochemical water quality of the lake in terms of hydrochemical parameters and micropollutants.

Due to the discharge of two wastewater treatment plants into the lake, persistent compounds can be expected to accumulate as it has been observed for European water bodies. Population growth will increase the pollutant load received by the lake and the natural self-purification of the lake will probably not be sufficient to eliminate these recalcitrant substances. Therefore, long-term monitoring of emerging polar **micropollutants** is strongly advisable, in particular since this has not been done to date. The set-up of the risk matrix based on concentrations measured and frequency of occurrence of organic micropollutants provides the basis for decisions in view of the refinement of future monitoring campaigns. Specific monitoring is also recommended for the existing drinking water treatment plants since many polar micropollutants are not removed in conventional drinking water treatment. In addition, detail planning should be updated frequently in order to adapt treatment technologies to futures changes in raw water quality. Concentrations of a number of micropollutants were below the detection limits or were found as much lower than in similar European or North American lakes. For most compounds the concentrations were below 100 ng L^{-1}. However, an additional fourth step in wastewater treatment using activated carbon is recommended to reduce micropollutant input into Lake Paranoá.

For **pesticides**, atrazine and its metabolites as well as diuron have been detected. While atrazine and its metabolites occurred in very low concentrations, most likely due to leaching from residues in soils or groundwater, diuron showed higher concentrations and high variability of concentrations. High proportions of pesticides may originate from (public) green spaces, house facades and so on. Therefore, WWTP effluents should be monitored as well.

DOC concentrations in Lake Paranoá are fairly low. Nevertheless, taking into account expected higher frequencies and intensities of extreme events, as a result of climate change as well as of urbanization and population growth, a regular monitoring of DOC concentrations or, at least, UV absorption at 254 nm is recommended. Applying advanced DOC-analyses, that is, SEC-UV/DOC analyses, trends of water quality changes can be noticed early in time and enable proactive action.

None of the **heavy metals** measured in lake water exceeded Brazilian and German drinking water standards, which emphasizes the high raw water quality. Sediments are not contaminated except for Cd and Pb but showed only a very low remobilization potential. The concentrations of **total phosphorus** and **chlorophyll-A** of the last years indicate that Lake Paranoá has reached a steady oligotrophic state. However, population growth and climate change cause serious concern about the maintenance of this state. Hence, future trends should be closely and carefully monitored to be able to react early to unwanted developments. With regard to future drinking water production it is a must to protect the raw water quality strictly. This can be done by minimizing the diffuse and direct input from land use and by optimizing wastewater treatment rather than by removing contaminats during drinking water treatment. A potential contamination of lake water with (emerging) substances or transformation products, which possibly elude the processes of drinking water treatment, could develop into serious threat to healthy water supply.

In view of further research, additional analyses are recommended to quantify the input via the tributaries and urban runoff during the rainy season. For example, using passive sampling devices can help to obtain reliable mass balances for the constituents of Lake Paranoá water. Together with a deeper understanding of the limnologic features, advanced modeling approaches can contribute to identify and understand the underlying processes in the lake as well as to avoid further possible sources of organic micropollutants. A refined model can also be used to simulate different climate change influences and population growth scenarios leading to more precise recommendations and thus to support successfully an integrated water resources management.

10.3 TECHNOLOGY

Regarding the surface water used for drinking water production in Brasilia, strong seasonal variations in rainfall are the major cause for extreme differences in raw water quality. These seasonal changes in raw water quality in turn result in variable treatment effectiveness. For the same treatment facility, differences in one order of magnitude of turbidity removal effectiveness between wet and dry seasons are typical. In general, simultaneous and optimal removal of both particles and dissolved organic matter remain as a challenge.

Out of the lessons learnt, the main recommendations for the future **drinking water treatment** plant for Lake Paranoá to provide high-quality drinking water should include (1) additional to the present drinking water treatment analytical and monitoring equipment to cope with natural fluctuations of particle concentrations in raw water and (2) the application of advanced treatment technologies to be considered as an alternative to conventional treatment technologies to enable the reliable removal of trace organic substances. This includes high-pressure membrane filtration (nanofiltration or reverse osmosis) as an add-on to a conventional train. Even more promising is a combination of coagulation/clarification and ultrafiltration with activated carbon adsorption. With such advanced technologies, as a long-term investment, meeting the increasing requirements for other quality parameters will be much easier in the future. Finally, it was also shown that the Distrito Federal has a promising potential for water savings both due to rationalizing the use of available water by increasing public awareness and by using rainwater to reduce potable water consumption for non-potable purposes.

A second focus of technology in the IWRM concept is on urban drainage/stormwater management and waste water treatment. General achievements of the project in terms of **wastewater treatment** are knowledge transfer on benchmarking methodology and advanced wastewater treatment techniques, for example, membrane technology. The experiments using a pilot plant for advanced wastewater treatment at the northern waste water treatment plant show a high efficiency of the applied methods for the removal of hazardous micro-substances. The next step towards the utilization of the lake as a drinking water source will be further research on other advanced treatment techniques and their large-scale implementation. In addition, a crucial point regarding waste water management in Brasília is the identification of incorrect and illegal connections and leakages of the sewer and drainage system with the aim to prevent the entry of polluted and non-treated water into the water body. Furthermore, avoiding the hydraulic overloads of the existing wastewater treatment plants, which might result in low quality of discharged water, is important.

The **stormwater management** in Brasília fulfills the conventional standards of a separate system. However, the quality of stormwater runoff draining into the lake and recent flood events has shown the need to reconsider the existing system in some aspects. It needs improvement by modern stormwater management techniques for flood and groundwater protection. In terms of water quality, runoff during the rainy season might be loaded with pollutants and sediment from surfaces without vegetation, for example, multilane roads or construction sites. A crucial part of stormwater management is the classification of source areas and runoff quality. The urban drainage system itself has to be considered as a crucial element of stormwater management in scope of an IWRM. In addition to measures aiming directly at the existing drainage system and flood protection, other measures might be considered. These might be percolation or decentralized or centralized retention. In order to optimize stormwater management the simulation of the complex interaction within the sewer systems, advanced simulations of surfaces and all other elements of urban drainage of the affected areas are recommended.

10.4 CAPACITY DEVELOPMENT

Capacity development has to be considered as an overall aspect within every IWRM approach. The main focus of capacity development was on the exchange of experiences from practice in Brazil and Germany. In addition, capacity development comprised also the exchange between the research partners of IWAS-ÁGUA DF and included lab courses, courses on modelling and others.

10.5 LESSONS LEARNED

The major outcome of the project IWAS-ÁGUA DF is a base of knowledge, and personal interrelations between scientists, practitioners and stakeholders in scope of an Integrated Water Resources Management (IWRM) for the Distrito Federal in

Central Western Brazil. An overall lesson learned during the project is that an approach linking knowledge, standards and demands of all participants from science and practice is required for the successful development and implementation of an IWRM concept. IWRM related challenges have a high complexity and a wide range of abstraction levels. Consequently, projects of the kind carried out and described here, need an approach balancing needs in practice and scientific standards. In this context, a crucial part of the project IWAS-ÁGUA DF was the consideration of economic development and the specific socio-economic structures. This included a close collaboration with local/regional stake-holders. In particular, we aimed at increasing expertise in all thematic fields through capacity development. The project IWAS-ÁGUA DF provides adapted solutions for an Integrated Water Resources Management in the study region in Central Western Brazil and thus, contributes to the achievement of the Millennium Development Goals. The results of the project might be transferred also to other regions in South America or worldwide with similar conditions.

IWA Publishing's authorised EU representative for General Product Safety Regulations is Diane D'Arras, 15 rue Duret, 75116 Paris, France, e-mail: safety@iwap.co.uk.

Printed and bound by CPI Group (UK) Ltd, Croydon, CR0 4YY

06/07/2026

02157564-0002